SERVICING ELECTRO

Volume 2, Part 2

Servicing Electronic Systems
Volume 2, Part 2

Television and Radio Reception Technology

**A Textbook for the
City and Guilds of London Institute Course No. 224
(as revised 1990)
and for
The Technician Education Council Level II Course in Electronics**

by
Ian R. Sinclair
B.Sc.
*Formerly Lecturer in Physics & Electronics
Braintree (Essex) College of Further Education*

and

Geoffrey E. Lewis
B.A., M.Sc., M.R.T.S., M.I.E.E.I.E.
*Formerly Senior Lecturer in Radio, Television and Electronics,
Canterbury College of Technology*

AVEBURY

Second edition published 1992 by
Avebury,
Ashgate Publishing Limited,
Gower House,
Croft Road,
Aldershot,
Hants GU11 3HR,
England

Ashgate Publishing Company,
Old Post Road,
Brookfield,
Vermont 05036,
USA

Reprinted 1996

A CIP catalogue record for this book is available from the British Library.

Printed and bound in Great Britain by
Biddles Ltd, Guildford and King's Lynn

ISBN 1 85628 810 2

Contents

Preface

As the end of the century approaches, the technology of electronics that was born in the twentieth century is by now the dominant technology in all aspects of our lives. The very nature of electronics has changed enormously in our lifetimes, from its beginnings in radio to its involvement in control of everything from food mixers to car engine performance, from games to industrial empires.

All of this makes the task of servicing electronic equipment more specialized, more demanding and more important. Servicing personnel play a very important part in maintaining the correct operation of a system. They not only need to develop a high level of diagnostic skills but they also need to be able to communicate their findings to others so that the reliability and testability of a system can be improved. This then may also demand the further skills required to modify an in-service system. In particular, the training of anyone who will specialize in servicing must be geared to the speed and nature of the changes that are continually taking place. Such training must include a sound knowledge of principles and the development of diagnostic skills, neither of which is likely to be superseded by any changes in technology. Another important factor is that with the increasing harmonization of technical standards in Europe, it is likely that knowledge of technical terms in several European languages will become an essential part of the training for servicing work.

Although this series of books is designed primarily to cover the most recent requirements of the City & Guilds of London Course No. 224 in Electronics Servicing, and also to provide coverage of the equivalent BTEC course, the books have been written mindful of the needs of the home-based distance learner. The approach is systems-based, viewing each electronic component or assembly as a device with known inputs and outputs. In this way, changes in technology do not require changes in the methods and principles of servicing, only to the everyday practical aspects which are continually changing in any case.

We have also taken every opportunity to look beyond the confines of the present syllabuses to the likely requirements of the future, and particularly to the impact of a single European market on both electronics and training. The books will invariably be amended in line with changes in the syllabuses and in the development of electronics, but the aim will be at all times to concentrate on the fundamentals of diagnosis and repair of whatever electronic equipment will require servicing in years to come.

A guide book has been prepared which contains useful course hints and comments on the questions included in the main text. This booklet, which may be freely photocopied, is available free of charge to lecturers and instructors from:

Customer Services,
Avebury Technical,
Gower Publishing Co. Ltd,
Gower House,
Croft Road,
Aldershot,
Hampshire, GU11 3HR.

The final book in the *Servicing Electronic Systems* series that is in the course of preparation is:

Volume 2 Part 3 *Servicing Electronic Systems* (Control System Technology)

Acknowledgements

The authors gratefully acknowledge the permission of the City and Guilds of London Institute and the Electronics Examination Board to reproduce extracts from the Course 224, Electronics Servicing (new scheme) syllabus and regulations. An abridged version of the syllabus is given in Appendix 3, and used as a cross-reference to the contents of the various chapters. For precise details of the scheme, the reader is referred to the full Part II syllabus (CGJ 30142 D13(2)) available from the City and Guilds of London Institute, 76 Portland Place, London W1N 4AA.

Grateful thanks are also extended to Philips Components Ltd, Thorn EMI Ferguson Ltd, SGS Thompson Ltd and to the Editor and the many contributors of the IPC magazine *Television*, for permission to use the diagrams and technical information extracted from their various publications.

In addition, the authors would like formally to recognize the contributions of the following course tutors who made many useful and constructive criticisms of the previous series, *Electronics for the Service Engineer:* Steve Dennis, Brockenhurst College; M. Diplock, Eastbourne College of Advanced Technology; M. T. Dunn, Harlow College; N. R. Farrow, Norwich City College; A. Hynch, Weymouth College; Ian Oxenforth, North Lindsey College of Technology, Scunthorpe; D. A. Norman, East Herts College; Chris Parlett, Highlands College, Jersey; G. Rigby, Runshaw Tertiary College; John F. C. Smith, Cumbernauld College; John J.Stanton, W. R. Tuson College, Preston; B. Woodgate, South Downs College of Further Education, Havant.

Introduction

This book covers the option Television and Radio Reception Technology of the City & Guilds of London Institute, Course 224, Electronics Servicing at the Part II level. It thus complements Volume 2 Part 1 of this series, *Basic Principles and Circuits (Core Studies)*.

The general objectives of this book are to enable the student to:

- interpret manufacturers' data and select equivalent components, including ICs;
- read and interpret block, circuit and layout diagrams;
- diagnose simple single component faults;
- identify the effects of faulty blocks in a system;
- set up a TV receiver for correct operation;
- adjust specified controls to correct faults.

With particular regard to TV receiver servicing, the importance of safe working practice cannot be too strongly emphasized.

In most of the TV receivers on sale in the UK, the metalwork of the chassis is connected to the neutral supply line (assuming that the plug is correctly wired). When any form of servicing is to be carried out to a working receiver, the latter must be supplied from an isolating transformer. Such a transformer has a 1:1 ratio and is rated to take the full power of the receiver, but with full electrical isolation between its primary and secondary windings. When a receiver is supplied in this way, the chassis can be safely earthed, since it is no longer directly connected to either of the mains supply lines. This is particularly important when using an earthed instrument such as a CRO for service work.

A further possible hazard lies in the practice of supplying the power for many modern colour TV receivers from a bridge rectifier, directly connected to the mains. This places the receiver chassis metalwork at half mains a.c. potential.

The service engineer should be aware of the implications of the Electromagnetic Compatibility/Interference (EMC/EMI) directive EMC 89/336/EEC. From the 1st January 1996, all industrial and domestic electronic or electrical equipment must carry the CE mark to signify that it meets these requirements. It is incumbent on the service engineer to maintain these standards during and after repairs.

Service engineers repairing TV receivers require ready access to a comprehensive collection of manufacturers' service sheets. They would be wise to keep their basic servicing knowledge up to date by the regular reading of such periodicals as *Television*.

1   Radio and television receivers

Summary

Aerials (antennae). Electromagnetic radiation. Receivers and interference. Tuner unit. IF stages. Demodulators.

Aerials (antennae)

The purpose of the receiving aerial is to intercept some of the energy of the modulated wave radiated by the transmitting aerial. Wave theory indicates that this can be done most efficiently by an aerial whose length is any exact multiple of half wavelengths of the signal itself. The smallest effective length of an aerial is thus one half wavelength of the desired signal. Since the wavelength, in metres, of the signal is equal to

$$\frac{3 \times 10^8}{\text{Frequency in Hz}}, \text{ or } \frac{300}{\text{Frequency in MHz}},$$

the length in metres of an aerial is given approximately by the expression:

$$\frac{150}{\text{Frequency in MHz}}$$

(for a half wave aerial). Though approximate, this is close enough for many

1

purposes. A small error occurs because 3×10^8 m/s is actually the velocity of electromagnetic waves in free space. The velocity in the metal conductors of the aerial is somewhat less.

The most common aerial element is the *dipole* shown in Figure 1.1(a), consist-

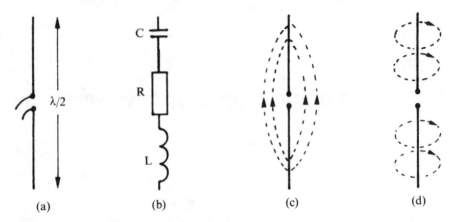

(a) (b) (c) (d)

Figure 1.1   A half-wave dipole and its characteristics: (a) half-wave dipole aerial; (b) equivalent circuit; (c) the electric field; (d) the electromagnetic field

ing of two equal-length rods connected to the transmitter or receiver via a feeder cable to the midpoint. The total length is approximately a half wavelength. Consider such a dipole to be connected to a transmitter. If the frequency matches that of the dipole, then the dipole behaves as a resonant tuned circuit as shown in Figure 1.1(b) with an impedance that is completely resistive. At frequencies on either side of this, the dipole acts either as an inductive or capacitive load to the transmitter, and a mismatch occurs. The resistance R of the equivalent circuit is usually called the *radiation resistance* for the aerial. R is not a physical resistance, only an equivalent one. If the aerial were replaced by a physical resistance of the same value it would dissipate the same amount of heat energy as the aerial radiates electromagnetic energy into space.

The a.c. voltage fed to the aerial causes each rod to take up opposite alternating polarities which give rise to the lines of force known as the *electric field*, shown in Figure 1.1(c). In a similar way, the a.c. current also produces alternating magnetic fields around the rods that generate the lines of force known as the *electromagnetic field*, shown in Figure 1.1(d). These forces act at right angles to each other and together cause the energy to be radiated away from the dipole. Figure 1.2(a) shows an electromagnetic wavefront approaching an observer, while Figure 1.2(b) depicts the mutual relationship between the two components of the electromagnetic radiation and the direction of the propagation.

If a complementary receiving dipole is to collect the maximum signal from the

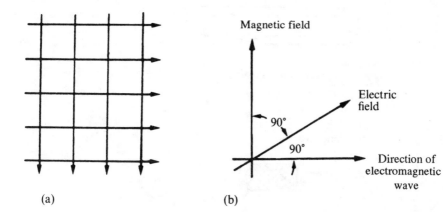

(a)   (b)

Figure 1.2   Electromagnetic waves in space: (a) wave front; (b) mutual relationship between components of propagation

radiated energy, then its conducting rods must cut the lines of the magnetic field at right angles. That is, the receiving dipole must occupy the same plane as the electric field of the radiation. If the angle is less than 90°, the induced signal voltage will be reduced proportionally. For instance, if the relative angle is 45°, the signal is 0.707 of maximum (3 dB down). The direction of the electric field of the wavefront is referred to as the *plane of polarization* of the wave. As shown above, this occupies the same plane as the transmitting and receiving dipoles. At VHF and UHF, vertical, slant and horizontal polarizations are in common use.

The angle of polarization of the transmitted wave can be changed somewhat, however, if the signal is reflected from some substantial object like a hill, a gasholder, or a block of buildings; and in some difficult reception areas a small change in the polarization angle to which the receiving aerial is set can give a significantly better signal level. Similar polarization adjustments can also be used, in the right conditions, to provide an additional way of discriminating against unwanted signals.

The radiation or *feed-point* resistance of the simple dipole is typically 75 Ω. Up to a point, an increase in rod diameter will increase the bandwidth over which the aerial gives acceptable reception. The overall length of the dipole is normally cut for a frequency lying in the middle of the band of signals for which operation is wanted. 'Folding' the dipole as shown in Figure 1.3(a) gives certain mechanical advantages but increases the radiation resistance to around 300 Ω. This figure can be reduced somewhat by increasing the cross-sectional area of the conductors.

A single dipole on its own seldom gives satisfactory signal pick-up unless the receiver is situated very close to the transmitter. By adding to the aerial a *reflector* situated *behind* the dipole and a number of *director* rods situated *in*

3

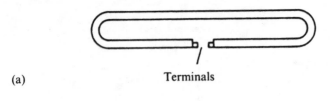

(a)   Terminals

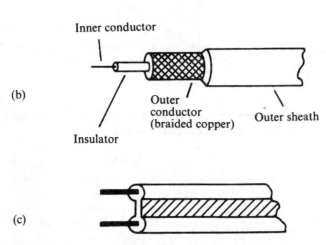

(b)

Figure 1.3   **Aerial components: (a) the folded dipole; (b) a cut-away section of co-axial cable; (c) a section of twin-line cable**

front of it in relation to the direction of the incoming signal, a much larger signal amplitude can be gathered. The aerial then becomes more highly *directional* (it has *high directivity*), accepting only signals arriving from a limited angle in front of the array (i.e. from the direction in which the director rods lie). The addition of these extra elements lowers the aerial's radiation resistance to nominally 75 Ω. Such an aerial (Figure 1.4) is termed a *yagi*, and is the usual form of TV aerial for a fixed site. The design of a yagi array is quite complex. Not only does it have to have high directivity, a definite value of feed-point resistance and a high gain relative to a dipole, but these parameters have to be maintained over a sufficiently wide bandwidth. Not unexpectedly, cheap aerials often give very poor results. The factors which make all the difference are often not apparent to the untrained eye (will that insulating material, for example, still be an insulator at 800 MHz on a wet day?). Aerials should always be bought from a reputable manufacturer who can specify and measure the aerial's performance.

If transmissions are being beamed from a single fixed site, the yagi has to be properly mounted so that it is pointing directly at this transmitter. This can be done approximately using a compass bearing, or by using a field-strength meter connected to the aerial. The receiver itself cannot be used as a reliable guide to

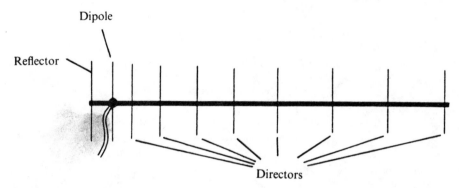

Figure 1.4   A yagi aerial

the received field strength unless its AGC line is first shorted, because AGC action will itself serve to mask the effects of signal variations.

Where two differently sited transmitters can be received by a single aerial array, two yagis can be connected to the receiver through a *diplexer* (a signal combiner) obtainable from the aerial manufacturer. Where several transmitters can be received from different directions (as happens in many parts of the USA), a rotating aerial assembly is often the only solution.

The feeder cables are designed to carry the signal with the least possible attenuation. Two common types of feeder are the twin-line (Figure 1.3(c)), which is used extensively in the USA and has a 300 Ω impedance, and the co-axial cable (Figure 1.3(b)) which is used almost exclusively in the United Kingdom and parts of Europe, and has a 75 Ω impedance.

It is essential that the impedances of aerial, feeder and receiver input are correctly matched. An aerial system using 75 Ω co-axial cable, for example, must feed into a 75 Ω matching aerial and a 75 Ω receiver input.

Twin-line feeders are described as being *balanced*. With neither line earthed, the signals in the two lines are always anti-phase and of equal amplitude. On the other hand, co-axial cable is said to be *unbalanced*, the earthed outer conductor acting as a shield for the signal-carrying inner conductor. A form of transformer called a *balun* (balance to unbalance) is available (Figure 1.5) to match one cable system to the other.

Exercise 1.1

(a)   Find out what transmitter frequencies are used locally. From the information sheets provided by the transmitting authorities (which in the UK are the BBC and the ITC (Independent Television Commission)) find also the radiated power of the transmitters and the size and contours of the service area.

(b)   Find out why the term *effective radiated power* (ERP) is used.

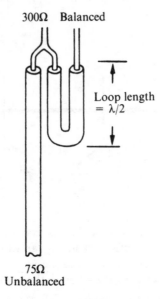

300Ω Balanced

Loop length
= λ/2

75Ω
Unbalanced

Figure 1.5   A balun capable of giving a 4:1 impedance conversion

(c)   Find out why all field strength maps are contoured in terms of microvolts per metre (μV/m).

(d)   If an aerial rotator is available, plot the polar diagram for the system for the lowest and highest frequency local transmitters.

Receivers

The requirements of any receiver can be summarized as follows:

1   to select the wanted frequency and reject all the unwanted signals that are present in any waveband;
2   to recover the information from the modulated wave;
3   to reproduce the information in a suitable manner, e.g. audio, vision, printer;
4   to carry out these functions without adding too much noise to the wanted signal during processing. Since noise power is proportional to bandwidth, the receiver bandwidth should be just wide enough to accommodate the wanted signal.

These requirements can best be met by using a receiver operating on the superhet principle as explained in Volume 1 of this series. Before studying this type of receiver further, it is instructive to investigate the forms of interference that can arise. Some of these are due to the use of the superhet, whilst others would occur no matter what type of receiver were used. The mixer principle is shown in Figure 1.6, where the two input frequencies f_1 and f_2 are distorted

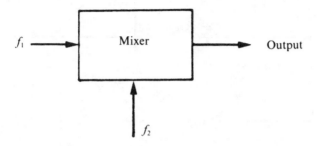

Figure 1.6   The mixer stage

together in the non-linear device. Amongst the frequency components present at the output are the original frequencies f_1 and f_2 and sum $(f_1 + f_2)$ and difference $(f_1 - f_2)$ frequencies. Normally, it is the difference frequency that is extracted using tuned circuits as the *intermediate frequency* (IF).

Interference and choice of IF

Transmitter channel spacings are allocated by international agreement to produce the minimum interference between adjacent channels. The actual channel spacing depends on the type of service to be provided. They vary from 9 to 10 kHz for AM sound broadcasting in the long and medium wavebands to 8 MHz for TV broadcasting in the UHF bands. Figure 1.7(a) indicates how *adjacent-channel interference* can arise from poor receiver selectivity. The effect can be minimized by using highly selective, low-frequency IF amplifier stages. Figure 1.7(b) depicts how *image* or *second-channel* interference arises. Either of the two frequencies f_a and f_b will beat with the local oscillator (LO) frequency f_0 to produce a difference frequency equal to the IF. Only one of these is the wanted signal. The other is the image or second-channel interfering signal. From Figure 1.7(b), it will be seen that the wanted and image frequencies are separated by twice the IF. Image-frequency rejection is therefore improved by using a high value IF, and by ensuring enough RF amplification and selectivity ahead of the mixer stage. Any strong signal present at the intermediate frequency can also break into the IF stages and be processed as an interfering signal. IF breakthrough, as this is called, can be minimized by adequate shielding and screening of the receiver, but more particularly by adopting a little-used transmitting frequency for the IF. This again is a matter for international agreement. From the foregoing, it can be seen that the choice of IF is a

7

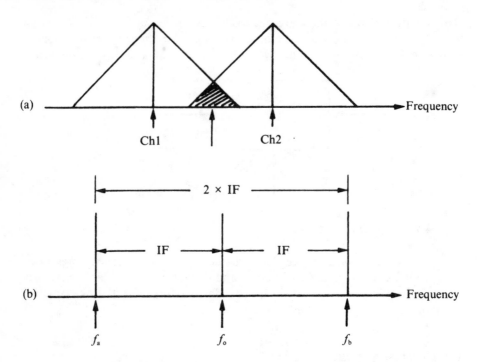

Figure 1.7 **(a) Adjacent-channel interference; (b) image or second-channel interference**

compromise, but one more factor is important: the value chosen must be high enough to accommodate the bandwidth of the signal to be processed. Some typical IF values and their areas of application are shown in Table 1.1.

Table 1.1

Radio frequency range	Waveband	Service	IF value	Base bandwidth	Channel bandwidth
150–300 kHz	Long	AM broadcast	465/470 kHz	5 kHz	10 kHz
525–1605 kHz	Medium	AM broadcast	465/470 kHz	5 kHz	10 kHz
88–108 MHz	VHF	FM broadcast	10.7 MHz	15 kHz	250 kHz
470–575 MHz (iv)	UHF	TV broadcast	⎧39.5 MHz	5.5 MHz vision	*
615–850 MHz (v)	UHF	TV broadcast	⎨33.5 MHz	15 kHz sound	*

*Both sound and vision signals are contained within the same channel bandwidth of 8 MHz. The gap between Band iv and Band v, 575–615 MHz, will be used for the new Channel 5 service. The digital NICAM stereo signal is carried on a sub-carrier at 6.552 MHz.

The AM tuner section (front end)

A typical AM radio 'front end' is shown in the circuit of Figure 1.8. The RF

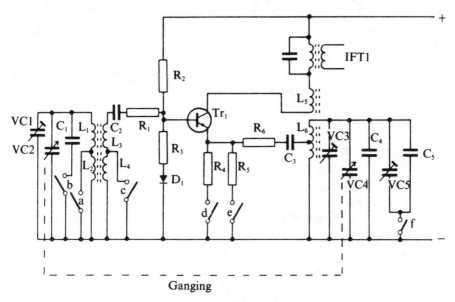

Figure 1.8 **The tuner section (oscillator-mixer) of an AM radio**

tuning coils are wound on to the ferrite rod which is used as an aerial.

The wave-change switch in this circuit comprises six separate sections, numbered (a) to (f) in the circuit diagram. For long-wave (LW) reception, switches (b), (d) and (f) are closed. For medium-wave (MW) reception, switches (a), (c) and (e) are closed.

Switch (a) connects in an extra portion of coil L_2 in the long-wave (LW) position of the switch and shorts out the extra portion (so as to avoid unwanted resonances) in the medium wave (MW) position. Similarly switch (b) connects in the additional tuning capacitor C_1 for LW use.

The signal selected by the resonance of the primary windings of L_1 and the tuning capacitor VC2 is coupled by the secondary winding through C_2. VC1 acts as a trimmer to adjust the range of VC2. The small-value resistor R_1, connected in series, is there to suppress any oscillations which might be caused by stray feedback.

At this stage, also, an additional section of coil L_4 is used for LW, but is shorted out by switch (c) for MW use.

Transistor input is conventional, and Tr_1 acts as a self-oscillating mixer, using feedback from collector to emitter. L_5 is the coupling winding for the oscillator, and L_6 the tuned winding, with VC3 and VC4 as trimmer and tuner respectively.

9

For LW use, the additional trimmer VC5 and fixed capacitor C_5 are switched in or out by switch (f).

The oscillator circuit uses a tap on L_6 to feed back signal through C_3 and R_6 (there to suppress unwanted oscillations) to the emitter of Tr_1. Switches (d) and (e) are used to connect in different emitter load resistors R_4 and R_5 for LW and MW respectively.

With Tr_1 oscillating, the input signal is mixed at the base-to-emitter junction with the oscillator frequency; and the intermediate frequency of 470 kHz so obtained is selected by the tuned load of the intermediate frequency transformer IFT1.

The FM tuner (front end)

The 'front end' circuit of an FM receiver is more complex than that of an AM receiver because of the much higher frequency of the signals. For instance, the simple self-oscillating mixer circuit always allows a fairly large amount of the oscillator signal to appear at the base of the mixer transistor. In AM radios this is not serious, because the oscillator frequency is 470 kHz higher than the signal frequency. Even at the highest MW frequency (which is around 1.6 MHz) oscillator frequency is still about 30% higher than signal frequency, and so is easily rejected by the tuned circuits at the input.

At the Band II frequencies (88 to 108 MHz) that are used for FM broadcasting, however, the standard IF of 10.7 MHz makes the ratio much less favourable, with oscillator frequency only some 10% higher than signal frequency. In addition, sharp tuning is much more difficult to obtain at high frequencies. Rejection of the oscillator frequency by the input tuned circuits becomes inadequate, and the oscillator signal is radiated from the aerial, causing interference to other FM users. FM receivers therefore need separate RF amplifier stages to minimize such radiation, and also to amplify and tune the input signal before mixing. Some designs even employ separate RF, mixer and oscillator stages.

A typical FM circuit of fairly old design has been chosen for illustration in Figure 1.9 so that the main points can be more easily brought out. A few decoupling capacitors have been omitted in the interests of clarity.

The signal from the 300Ω aerial line is coupled by the input transformer T_1, whose secondary is tuned by C_1 to a frequency in the centre of the FM band. No adjustable tuning is used in this part of the circuit because the tuned circuit is so loaded by the emitter circuit of Tr_1 that the bandwidth is very large.

Owing to the use of the bypass capacitor C_9, both power supply rails are at the same signal potential. Therefore the collector load effectively consists of the parallel combination of L_1, the main tuning capacitor VC1 and the trimmer capacitor CT1. The tuning range is typically 88–108 MHz.

Tr_2 is operated as a self-oscillating mixer, with the oscillator tuned circuit

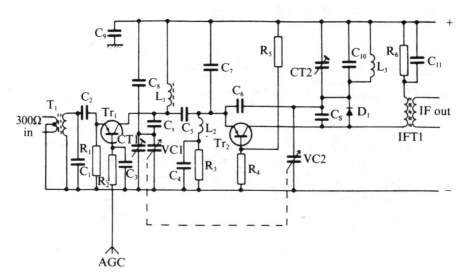

Figure 1.9   Simple FM tuner circuit

(L_3–VC2–CT2) feeding back signal in phase to the emitter through the low-value (only 1 to 5 pF) C_6.

The IF is selected by the transformer IFT1, whose secondary is tuned to the standard value of 10.7 MHz.

The diode D_1 is an overload diode used to limit the amplitude of oscillation, and also to provide enough loading to prevent any oscillation at 10.7 MHz.

L_2, C_4 have negligible series reactance, allowing R_3 to be used as an earth return for the emitter.

Capacitors C_S have negligible reactance at these frequencies.

Refinements to the FM front end

The FM *front-end* circuit shown in Figure 1.9 used common-base transistor connections because of the better frequency response of a transistor connected in this way. A more efficient tuner that uses dual-gate MOSFETs in the RF amplifier and mixer stages, is shown in Figure 1.10. A major advantage of using MOSFETs is that their input impedance, even at 90 MHz, is high, so that damping of the tuned circuits is negligible and the input to the RF stage can therefore be variable-tuned to improve adjacant channel rejection.

Tuning is also assisted by the use of *varicap* (or *varactor*) diodes, which are semiconductor diodes connected as part of the capacitor in a tuned circuit, and which are reverse-biased. Since the capacitance between the leads of a varicap diode (indeed, of any junction diode) varies as the reverse bias varies, circuits can be tuned by adjusting the d.c. voltage bias on the diode rather than by

11

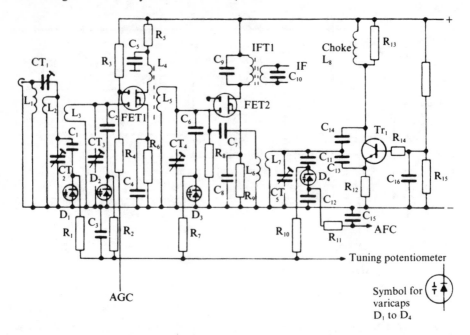

Figure 1.10   A MOSFET FM tuner circuit

mechanical movement. Large numbers of varicaps can be 'ganged' simply by using the same bias supply for each of them, and the absence of moving parts in the tuner greatly reduces unwanted feedback problems caused by stray capacitance.

In the circuit of Figure 1.10 diodes, D_1, D_2, D_3 and D_4 are varicaps. D_1 is used to tune the FM input stages, and D_2 to tune the signal input to the RF amplifier, FET1. The second gate of this FET, which controls gain, is used to apply AGC voltage; the dual-gate construction ensures excellent isolation of the AGC line from the signal circuits.

The signal is then passed by L_4 and L_5 to one gate of FET2, whose second gate is used to inject the oscillator voltage from Tr_1 — an ordinary Colpitts oscillator which is tuned by the varicap D_4.

The non-linear characteristic of the MOSFET ensures good mixing, with a good proportion of the wanted 10.7 MHz signal present in the output. This signal is taken from the drain of FET2 by the IF transformer IFT1.

The advantages of using dual-gate MOSFETs, then, are as follows.

1   They have high input impedance, so that damping is reduced. The selectivity of tuned circuits is improved and interference from other signals thus reduced.
2   Isolation between the two input gates is good, so that unwanted feedback can be minimized.

3   The characteristic of a MOSFET has a shape ideal for use as a mixer, so that a better ratio of IF signal output to VHF signal amplitude input is obtained.
4   Better signal-to-noise ratio can be obtained at high frequencies than is possible when bipolar transistors are used.

The advantages of using varicap diodes are as follows.

1   Since tuning is done electrically by d.c., the tuning control can be a potentiometer which will fit in anywhere in the circuit layout.
2   No moving parts need be used anywhere in the tuner.
3   Large numbers of varicap diodes can be 'ganged', so that it becomes possible to use more variable tuned circuits than can be done if ganging is mechanical.
4   It produces a smaller tuner unit.

TV tuners

TV tuners can be classified by the methods used for channel selection. The methods are as follows:

1   variable capacitance, using ganged capacitors;
2   variable inductance, using moveable cores which are also ganged;
3   varicap tuning, using special reverse-biased diodes whose self-capacitance is varied by altering the reverse-bias voltage.

A typical tuner unit is shown in Figure 1.11. The low-impedance coaxial cable from the antenna feeds into the emitter of a common base RF amplifier Tr_1, whose output in turn feeds the signal from a tuned line into the input of the second RF amplifier Tr_2, also operating the common base mode. The output from Tr_2 is coupled by an inductive link (L_4, L_5) to the tuned input of the mixer Tr_3. Tr_4 oscillates due to positive feedback provided by the capacitor between emitter and collector at a frequency controlled by the line L_9 and associated capacitors. The tuned circuit of the oscillator is coupled to the emitter of Tr_3 together with the RF signal. The IF is taken from the collector of Tr_3 into a normal IF transformer.

In the circuit shown, inductors acting as IF traps and chokes are indicated by the conventional inductor symbol. The components L_2, L_4, L_5, L_7, L_8 and L_{10} are coupling links, with L_{10} short-circuited so as to introduce damping and *reduce* the inductance of L_9 and so enable the oscillator to operate at a frequency higher than that of the incoming signal.

The remaining inductive components, L_1, L_3, L_6 and L_9, are the main tuning inductors and are shown as thin metal bars. The reason for this effect can be

13

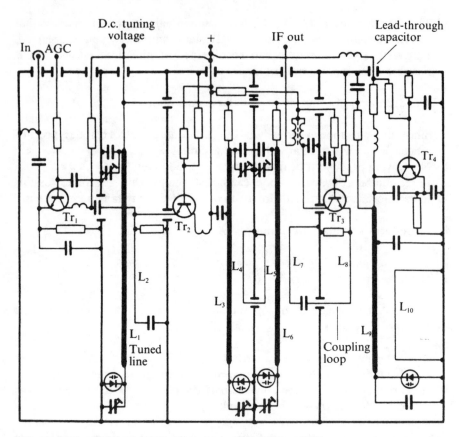

Figure 1.11 Typical varicap/resonant-line tuner unit

explained as follows. Since the reasonably attainable capacitance values in such a circuit as this cannot possibly be lower than the stray capacitance of the circuit itself, the inductance of a short piece of straight wire is enough to tune a circuit to frequencies in Bands IV and V.

At such frequencies as these, current flows mainly along the *outside surface* of conductors (skin effect), so that thick silver-plated wires are used. A good many of the accepted circuit laws of electronics need to be modified to take account of this behaviour at UHF.

Since it is vital to prevent unwanted feedback from one stage to another, each section of the tuner needs to be carefully screened from its neighbours. The mechanical construction of a UHF tuner is at least as important as its electrical design. Never attempt to open up a tuner unless a manufacturer's service sheet is available and suitable equipment is to hand for re-aligning the tuning.

Servicing is for the most part confined to taking measurements of the IF signal voltages and the d.c. supplies, and comparing the results with the manufacturer's figures. Large differences between the expected values and the readings actually obtained indicate fault conditions which can normally only be remedied by replacement of the complete tuner. Current production tuner units are constructed on a printed circuit board of alumina, and use surface-mounted components, resulting in a very small, highly efficient unit at relatively low cost. This makes it uneconomic to attempt any repairs except to simple faults.

Figure 1.11 also shows how an AGC control voltage is applied to the base of the first stage of RF amplication in order to avoid overloading both this and subsequent stages on large input signals.

The IF stages

This section forms the heart of the receiver. for it is here that most of the gain and selectivity takes place. Except for a few specialized cases, the IF frequencies for an AM receiver are typically either 465 or 470 kHz, with a total bandwidth of about 10 kHz. The FM receiver uses an IF of 10.7 MHz, with a bandwidth of about 200 kHz for *mono* and about 250 kHz for *stereo*. Although for VHF FM broadcasting, the frequency deviation maximum is limited to $\pm$ 75 kHz, the bandwidth of the signal is greater than 150 kHz because of the distorted waveshape of the modulated signal.

The voltage gain of a complete IF stage is large — of the order of a thousand times or more in a sensitive receiver — and the number of IF stages used will determine the amount of overall gain.

The narrow bandwidth and comparatively low frequency of the IF in the AM receiver permit the use of very sharply tuned circuits having large dynamic resistance values. Adequate IF gain can therefore often be achieved with only one or two stages of transistor amplification. A typical circuit of a simple receiver is shown in Figure 1.12.

In this circuit, the primary windings of both IF transformers are tuned, and a single amplifying stage is used. The detector diode D_1 is permanently forward-biased by the resistor chain R_4-R_2-VR_1 to increase sensitivity. The bias for the base of Tr_1 is also derived from this same resistor chain.

When a large-amplitude signal is demodulated, there will be a steady negative voltage at the anode of D_1 as a result of rectifying the carrier wave, and this will reduce the size of the bias voltage at the junction of R_4 and R_2, and also the bias on the base of Tr_1. This reduction in bias voltage will cause a lower steady current to flow in the transistor, so that its gain becomes less. Thus the circuit acts a simple but effective form of AGC.

Both IFT primaries in this small receiver are made to peak sharply at 470 kHz (see Figure 1.13) with the result that the IF response itself acts to reduce interference from adjacent transmissions.

In the MW band, each carrier is spaced at 9 or 10 kHz intervals, so that the

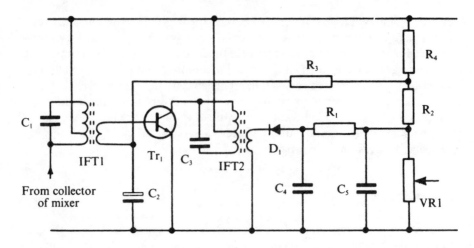

Figure 1.12   The IF stage of a pocket-size AM radio

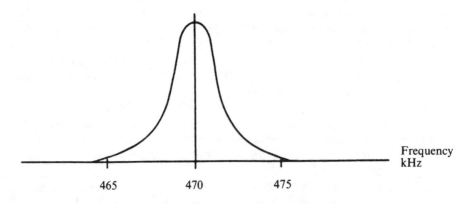

Figure 1.13   The ideal response curve for an AM receiver

amplitude of the IF response needs to be kept well down at a bandwidth of
5 kHz to avoid interference from the sidebands of adjacent channels.

Figure 1.14 shows a portion of the IF circuit of an FM-only receiver. In this
example, a single IFT (not shown) is used to couple the mixer stage to the first
IF amplifier stage. The coupling between the first and second IF amplifier
stages, and also that between the second IF to the next stage, is done by means
of ceramic filters called *transfilters*. These are in effect transducers which convert
the electrical IF signal into mechanical ultrasonic vibrations to which the
ceramic crystal resonates, and which then convert the vibrations back to electrical form again. By using these transfilters, it is possible to obtain a response

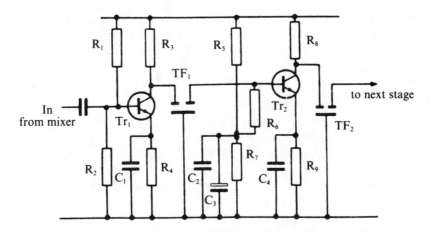

Figure 1.14   Portion of an IF circuit using transfilters

curve with much steeper sides and a flatter top. There is therefore better selectivity for signals, and less risk of oscillation being set up.

Figure 1.15 shows a portion of the IF circuits of a combined AM/FM receiver. The frequencies of the two IFs are so different that the two sets of IF transformers can be connected in series, though the 10.7 MHz IFTs need to be

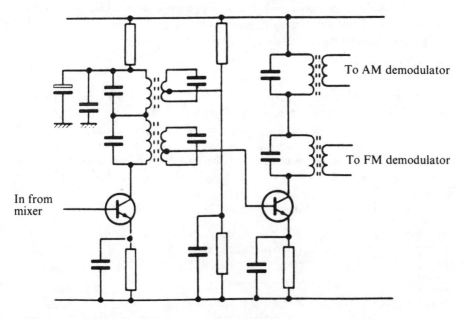

Figure 1.15   The IF stages of a combined FM/AM receiver

physically placed as close to the relevant transistors as possible. Damping resistors connected in parallel with the 10.7 MHz IFT windings may be used to control the bandwidth of the FM IF response.

Exercise 1.2

Align the IF stages of an AM receiver as follows. Disable the AGC system by shorting the line to earth. Connect a high-resistance d.c. voltmeter across the detector load. Couple the output of an AM signal generator to the mixer output connection via a capacitor of less than 100 pF. Set the signal generator level to a few tens of millivolts and adjust the attenuator to provide a just-readable deflection on voltmeter. Adjust the IF cores in turn to produce maximum meter deflection.

   Note. It might be necessary to reduce the level of output from the generator as the stages are brought into alignment.

   A better method of achieving a sharply-peaked IF response curve is to use a test instrument called a *wobbulator* or swept frequency signal generator. Indeed, its use is essential for the correct tuning of FM and other wideband IFs.

   This instrument is always used in conjunction with an oscilloscope from which a timebase waveform can be obtained. This timebase is used to control the frequency of the oscillator so that it varies from below the IF to above the IF as the timebase sweeps. If the output of the receiver is now displayed on the oscilloscope, the trace will be the graph of the frequency response of the receiver under test.

   Figure 1.16 shows the block diagram, and typical traces.

The TV IF strip

The *IF strip*, as its name suggests, comprises all those voltage-amplifier stages which work at intermediate frequency. The name derives from the appearance of the stages in early receivers, in which they were always built in the form of a long strip. The required IF response curve is shown in Figure 1.17. The amount of gain required over so wide a bandwidth can only be achieved by several stages of amplification some or all of which will in modern receivers be provided by ICs.

   The bandwidth required is achieved by:

(a)   loading each tuned circuit by connecting resistors in parallel to reduce the Q factor and increase the bandwidth, and/or by

(b)   tuning successive LC circuits to different frequencies in the IF range — a process known as *stagger-tuning*, or

(c)   using *surface acoustic wave filters*.

The circuit diagram of an early colour TV receiver IF strip shown in Figure 1.18 has been selected simply to show the principle involved.

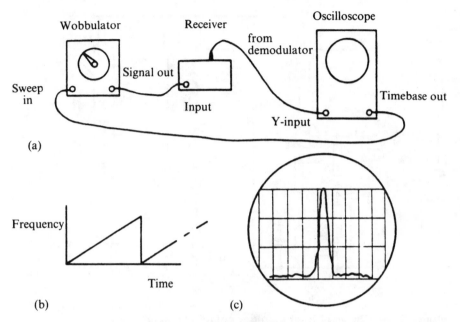

Figure 1.16 The wobbulator: (a) block diagram of arrangement; (b) time-base output; (c) typical oscilloscope trace

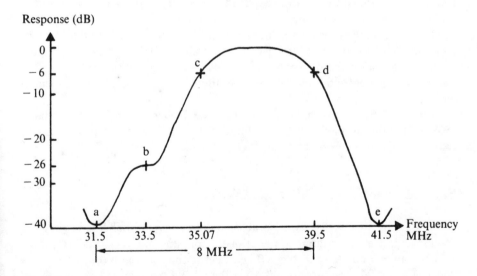

Figure 1.17 TV IF response curve: (a) adjacent vision channel; (b) sound carrier; (c) colour sub-carrier; (d) vision carrier; (e) adjacent sound channel

19

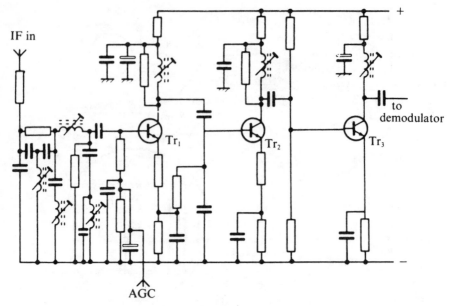

Figure 1.18   The IF strip of an older colour receiver

The inductors which are used as the collector loads of each of the three transistors are tuned by the stray capacitances of the circuit. AGC is applied to the base of the first transistor.

The feed from the tuner to the first IF stage also includes a set of filter circuits which operate as traps tuned to reject adjacent-channel signals (shown in Figure 1.17). These filters should never be readjusted during normal servicing. The settings are critical, and should only be altered if suitable test equipment is available.

More modern designs make much more use of filtering circuits, particularly when high-gain untuned ICs are used in the IF stages. Transfilters also help to reduce the number of ordinary tuned circuits used in the strip.

Surface acoustic wave filters (SAWF)

These devices are chiefly used because, unlike conventional filters, they are economical, generate negligible phase distortion, and are very small (typically 1.5 cm diameter and 0.3 cm thick). Their operation is based on the *piezoelectric effect* of a crystal structure. In this case the crystal material is usually lithium niobate with a surface area of about 10 mm × 4 mm. Two sets of interleaved fingers of aluminium form the input and output as shown in Figure 1.19. When a signal is applied, an alternating electric field is set up underneath the input

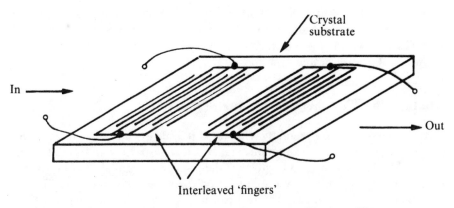

Figure 1.19   Basic construction of a surface acoustic wave filter

electrodes and this causes the crystal structure to distort also in an a.c. manner. This results in the propagation of a wave of energy along the surface, in a manner similar to acoustic waves travelling through solids. As the wave passes under the output electrodes, it generates an a.c. output voltage, provided that the geometry of the *finger* structure matches the wavelength of the signal. Used in this way, a single device can produce all the response curve shaping needed. An important service advantage is that the SAWF cannot be misaligned and does not therefore need to be realigned. The one minor disadvantage is the SAWF's *insertion loss* (attenuation due to being inserted in circuit), which is about 20 dB. Figure 1.20 shows a typical application of a SAWF in a modern

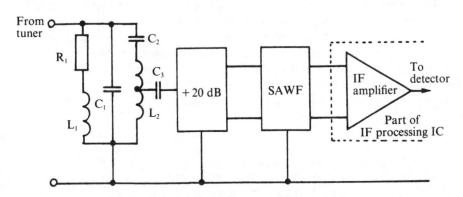

Figure 1.20   IF stage circuit using surface acoustic wave filter

TV receiver. A simple bandpass filter with a wide frequency response preselects the output from the tuner unit. The damping resistor R_1 increases the bandwidth and the tapped inductor L_2 is used to match the fairly high filter impedance on to the relatively low fixed-gain amplifier's input impedance. The gain

21

of this stage is fixed at about 20 dB to make up for the insertion loss of the SAWF, which has input and output impedances of about 1.5k Ω. The main IF amplification, which is gain-controlled, takes place in an IC that also contains its own AGC, detector and sync stages.

Faults in the TV tuner and IF stages

IF strips themselves are generally reliable, so it is always advisable to check the power supply and the video amplifier stages before suspecting an IF fault of any kind. A good rough check is to advance the *contrast* control of the receiver and watch for video noise ('snow') to appear on the screen. If it does, the IF strip must be operating correctly. The fact can be confirmed by injecting signals at intermediate frequency into the strip, whereupon a pattern should appear on the screen of the CRT.

Failure in the IF stage, when it occurs, is often due to an o/c capacitor, particularly a coupling or a bypass capacitor. Other possibilities include the AGC circuit or faulty transistors.

Unlike older tuner units containing mechanical switch contacts, modern tuner units are particularly reliable. A large noise signal, with no picture or sound, often indicates that the local oscillator has failed. A picture which has *suddenly* become noisy, however, can be caused by an o/c capacitor, or by poor connections in the aerial lead to the tuner. Most tuners of the sealed type incorporate test points at which the presence of bias or signal voltage can be checked.

The RF stage of any tuner is particularly liable to overload damage caused by the effects of distant lightning. Nothing at all can protect a tuner from the effects of a direct stroke of lightning, but the overload diodes and spark gaps used in modern tuners give excellent protection against the large interference pulses which are radiated during thunderstorms. Tuners of older design are not so well protected (some are not protected at all), so a receiver which has failed during a storm has probably suffered a burned-out RF stage.

Exercise 1.3

Use the wobbulator or swept-frequency technique described under Exercise 1.2 to display the IF response curve of a monochrome or colour TV receiver. Identify the position of the various frequencies shown in Figure 1.17.

Demodulation, AFC and AGC

The simplest method of demodulating an AM signal involves the use of a diode detector (rectifier). Often a germanium type is preferred for its lower series resistance. The AGC voltage is also derived from this diode, as is shown in Figure 1.21. An increased amplitude of input signal will result in a larger

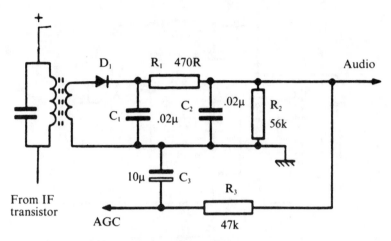

Figure 1.21   A typical AM demodulator stage

rectified voltage being developed across R_2. By filtering off the audio with R_3C_3, a d.c. gain-control voltage can be generated to reduce the gain of the IF stages and restore the output to its original level. ICs for AM demodulation often contain a synchronous detector. The process consists of multiplying (or beating) the incoming IF signal with the output from an oscillator running at exactly the carrier frequency. The modulation can then be extracted by using a simple low-pass filter. Such demodulators have the advantage that they can also be used for processing suppressed carrier forms of modulation.

Several types of demodulator are to be found in FM receivers. Portable receivers and low-cost tuners use the ratio detector, whose circuit is shown in Figure 1.22. This circuit is easily identified by the 'series' connection of its diodes.

Better-quality tuners will use the so-called Foster–Seeley discriminator circuit of Figure 1.23, in which the signal is fed directly from the final IF amplifier to the cathodes of both diodes (or to the anodes of both diodes, if the diodes are reversed).

The most modern receivers use ICs for demodulation. A favourite circuit is the so-called *phase-locked loop* (PLL) to recover the audio signals from the modulated IF. The action of the PLL circuit was described in Volume 1 but can be summarized as follows. The output of a voltage-controlled oscillator (VCO) is forced to follow the frequency variations of the FM input signal by an error voltage that has been derived from any difference between the two signals. The error signal will be found to represent the modulating component of the incoming signal.

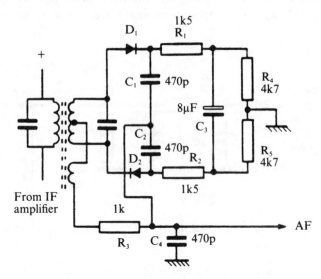

Figure 1.22　A ratio discriminator for use in a receiver

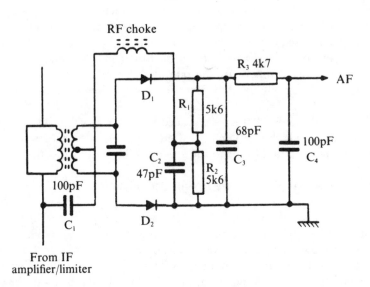

Figure 1.23　The Foster–Seeley FM discriminator

In general, the ratio detector requires fewer stages of IF amplification than does the Foster–Seeley type, but it has a higher distortion level. Both types need careful adjustment of the cores of the coils to give good performance, which explains the trend to the use of PLL circuits which have fewer critical settings. The way in which the other FM demodulator circuits operate is beyond the scope of this book.

Irrespective of what type of demodulator system is used, a graph of output voltage from a demodulator plotted against the IF input to it should ideally have the S shape shown in Figure 1.24. An actual graph can be quickly obtained

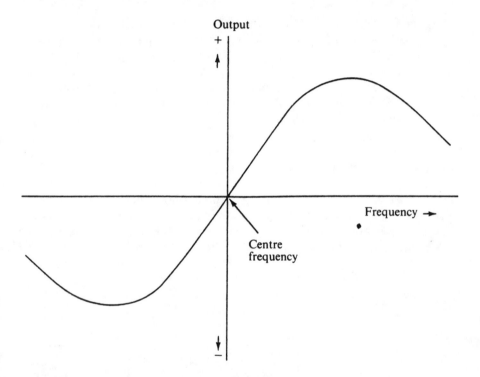

Figure 1.24   S-curved response of FM demodulator

by using a wobbulator and an oscilloscope. Irregularity in the shape of the S-curve may point to a fault in the discriminator, or to a dip in the IF response curve, the cause of which should next be investigated.

Automatic frequency control (AFC) is achieved by using a d.c. voltage derived from the signal source, either from the AGC line, where the voltage reaches a maximum value at the point of correct tuning, or (as is more common) from the audio signal, which develops a d.c. component, positive or negative, depending on the direction of the tuning error. However it is obtained, the AFC voltage can be used to bias the varicap diode used for oscillator tuning to correct the tuning error, as shown in Figure 1.10.

Note that some high-quality tuners use so-called *quieting* circuits, in which the output of the detector is cut off, or 'muted', until a sufficiently strong signal (measured by the AGC voltage) is received. These circuits need to be shorted out whenever any measurements of IF or demodulator response are being made.

Exercise 1.4

Drive the IF stages of an FM receiver with a wobbulator set to 10.7 MHz, and connect the *Y*-input of the oscilloscope to the IF output. Measure the bandwidth of the IF response. Note the effect of adjusting the positions of the IF transformer cores.

Connect the oscilloscope *Y*-input to the output of the demodulator and inspect the S-curve which should appear. Note the effect of varying the cores of the inductors in the demodulator.

Frequency synthesis tuning

Figure 1.25 shows the way digital signal processing is beginning to influence analogue designs; the diagram outlines the basic principles of such a system. The local oscillator signal is provided by the phase-locked loop (PLL) consisting of voltage-controlled oscillator, dividers, filter and comparator. The *prescaler*

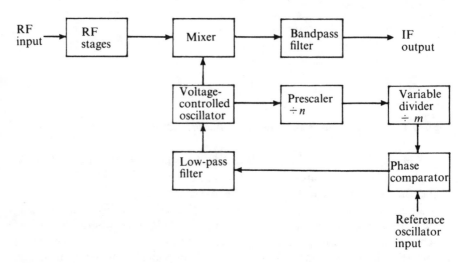

Figure 1.25 Frequency synthesis tuning

$(\div n)$ is used to extend the frequency range of operation. The basic reference frequency is provided from a crystal oscillator whose output is divided down to provide an input to the phase comparator.

An example from television gives a useful explanation of the operation. The reference frequency is provided from a 4 MHz crystal oscillator, the output of which is divided by 1024. The receiver local oscillator thus locks to multiples of 3.906 25 kHz. When a channel is selected, control logic converts this into a frequency, and the phase comparator causes the local oscillator to be adjusted until the output from the variable divider $(\div m)$ is equal in phase and frequency to 3.906 25 kHz. If the prescaler has divided the local oscillator frequency by 16, the actual tuning steps become $3.906\ 2 \times 16 = 62.5\text{kHz}$.

Thus by choosing suitable reference frequencies and division ratios, any tuning step size can be generated. Often for a specialized communications receiver, the step size is as small as 10 Hz, and this gives the operator the impression of almost continuous tuning. The output from the local oscillator can also be fed to a frequency counter that has been offset by the IF value. The counter output can then be used to provide a digital read-out of the tuned frequency, for display either on LEDs or as on-screen text. Because the division ratios are scaled in binary, it is quite easy to adapt this concept to digitally controlled tuning from a device such as a key pad.

A further digital extension provides for the *scanning receiver*. This type of receiver can be made to scan-tune a particular waveband automatically and stop when a signal with predetermined characteristics is found. Frequency synthesis tuning is very stable since all injection frequencies are obtained from the same high-stability crystal oscillator. Problems that can arise include noise in the form of jitter from the oscillator, and spurious beat notes generated from the high-speed synthesizer divider circuits.

Test questions

1 (a) Calculate the approximate length of a half-wave dipole element for an aerial operating at 600 MHz.
  (b) Explain why certain television aerials are mounted with their elements in the vertical plane.

2 For an FM receiver tuned to 92.3 MHz,
  (a) state the standard value of the intermediate frequency;
  (b) calculate the local oscillator frequency;
  (c) state the mixer output signals;
  (d) state the image frequency;
  (e) sketch the IF response curve and label the typical centre and band-edge frequencies.

3   Sketch the IF response curve for a typical UK PAL television receiver, showing the values of the wanted and adjacent sound and vision carrier frequencies.

4   For a UK PAL television system state the frequency spacing between:
    (a)   Channel 24 and Channel 25 vision carriers;
    (b)   Channel 24 sound carrier and Channel 25 vision carrier;
    (c)   the sound and vision carriers for Channel 24.

2   Television fundamentals (monochrome and colour)

Summary

TV fundamentals. Scanning and interlacing. Vestigial sideband. Colour triangles and colour mixing. Colour camera. Colour-difference signals. Modulation. Grey scale. Compatibility.

The first thing the reader should do before embarking on this and the following chapters is to refresh his/her memory of the block diagram of a simple colour TV receiver, shown here as Figure 2.1 and already discussed in some detail in Volume 1. The NTSC and SECAM receivers have a similar organization; the major differences are in the way in which the chrominance signals are processed.

Scanning and interlacing

All sounds consist of waves of air pressure; the only component theoretically needed to convert these waves of air pressure into electrical waves is a simple transducer — the microphone. A picture is not so simple to convert into a waveform. Although light is itself an electromagnetic wave, it has a frequency range which is too high to be reproduced by direct electronic methods, so some means is needed for converting the information which a picture conveys into a waveform of a frequency which can be more easily handled. That method is *scanning*.

Consider a small dot of light. It is possible to construct transducers capable of converting into voltage signals the amounts of brightness (*luminance*), of colour

29

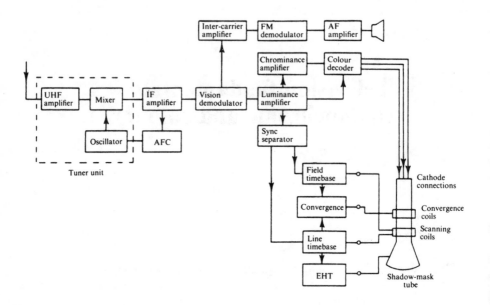

Figure 2.1   Simplified block diagram of a colour TV receiver

(*hue*) and of percentage of colour (*degree of saturation*) which such a spot of light contains. The camera tubes of a TV camera between them make up such a set of transducers.

Now just as a newspaper photograph, if you examine it closely, proves to be made up of tens of thousands of dots, so a TV picture can be made up of similar dots. The process of converting the picture into a waveform becomes much simpler, however, if the dots are sampled (very rapidly indeed) in a definite sequence, rather than just at random. This sampling process is called *scanning*.

The transducers used in studio TV cameras are a form of cathode-ray tube, and scanning is carried out by appropriate deflections of the electron beams of these CRTs, the size of dot sampled being equal to the diameter of the electron beam. Modern cameras tend to use solid-state *charge-coupled device* (CCD) image sensors, but the scanning process still applies. The scan pattern used consists of a series of horizontal lines across the picture (with the 'dots' along the lines being allowed to overlap), followed very rapidly by a second line scanned below the first, a third below the second, and so on until the whole picture has been scanned. In practice, for reasons which will be seen shortly, the horizontal lines of the observed picture are not scanned in strict numerical order. Instead, the odd-numbered lines — 1, 3, 5, 7, 9, etc. — are scanned in the first vertical sweep of the spot, and the even-numbered lines — 2, 4, 6, 8, 10, etc. — in the immediate succeeding sweep. So fast is the beam's speed of travel that the 'missing' lines are filled by the second scan before the human eye, which has a considerable degree of *persistence of vision*, has had time to notice their absence.

Two timebases are therefore required for the scanning beams. The first is a *line timebase* to govern the horizontal deflection of the scanning beam, and the second is a *field timebase* to govern its vertical deflection at the end of each completed field. The speed of the line timebase is, of course, much greater than that of the field timebase. In the United Kingdom, the ratio is set at 312.5:1.

A considerable number of horizontal lines per picture are needed if the picture is not to appear 'liney', showing its line structure too clearly. Moreover, in order to prevent the picture from appearing to flicker, a large number of complete vertical scans per second is needed. Motion-picture experience showed that the minimum was at least 24, and 25 is the number used in practice. Ideally, both the number of horizontal lines per frame and the number of complete pictures per second should be made as high as possible; but here the *bandwidth* of the resulting signal soon becomes a limiting factor.

The required bandwidth is calculated as follows. Assume the picture is broken down into square dots. In the United Kingdom TV system, 625 lines are used to form a *picture* (which is in turn composed of two fields or complete vertical scans of the screen). But only 575 of these 625 lines are used to convey actual picture information, the 'lost' 50 lines being used to provide flyback time for the vertical timebase, to transmit teletext data and to provide transmitter service information. A picture therefore requires 575 dots in the vertical direction: see Figure 2.2 in which are shown (a) picture dimensions and (b) the video waveform which would be produced by a scan of a simple alternating black and white pattern.

By international agreement, the transmitted width-to-height dimensions of the picture screen (the so-called *aspect ratio*) are always in the proportion of 4 : 3. The number of dots along a line should therefore be $575 \times 4/3 = 767$. Since the finest detail one could expect to see would be a bar pattern in which these dots were alternately black and white, that would require a waveform completing 767/2, or 383½, complete cycles in every line. With 575 lines per picture and 25 complete pictures per second, the bandwidth that would be needed is $767/2 \times 575 \times 25\,\text{Hz} = 5.513\,\text{MHz}$. The actual displayed aspect ratio of 5:4 is achieved by using a CRT mask that introduces about 7% of horizontal overscan.

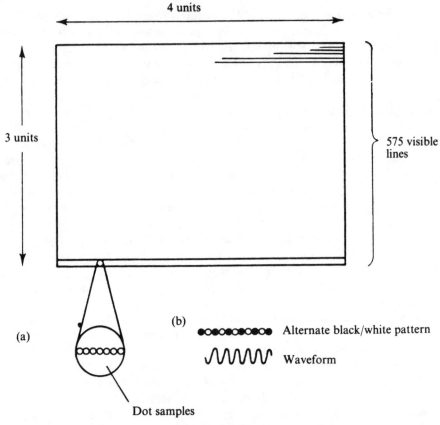

Figure 2.2 **(a) Picture dimensions; (b) the video waveform produced by a scan of a simple alternating black-and-white pattern**

To avoid flicker, the vertical timebase in Europe runs at 50 Hz (in the United States the figure is 60 Hz). If a TV system showed 50 complete *pictures* per second, however, the bandwidth required would be *twice* that calculated above, which would cause much difficulty in channel allocation and receiver design.

To reduce the required bandwidth, a technique called *interlacing* is used to enable a field timebase rate of 50 Hz to be used with a *picture repetition rate* (or *frame rate*) of only 25 Hz. As was briefly mentioned above, every vertical field is made to consist of only *half* the total number of picture lines, spaced over the whole vertical dimension of the picture. The principle as observed on the receiver screen is illustrated in Figure 2.3.

The first field traces out the odd-numbered lines, and the next field traces out the missing even-numbered lines. Two *fields* therefore make up one *frame* of the picture. (In practice, the transmitted lines of each frame are numbered in strict sequence from 1 to 625.)

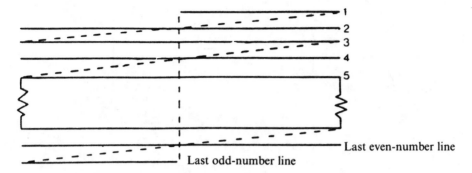

Figure 2.3   The observed principle of interlacing

It follows that every field must contain a half-line, either at the top or at the bottom of the picture. Interlace is then achieved automatically as soon as the speeds of the line and field scans are correctly synchronized.

Synchronization

The *luminance* (brightness) signal received from the TV camera is a waveform whose amplitude is proportional to the brightness of the picture at every dot scanned. For a receiver to be able to 're-draw' this picture on the viewer's screen, the timebases of the receiver must run at the same frequency as the timebases of the cameras. In other words, every line scan must start just before the first part of the line waveform, and every field scan must start at the correct line or half-line.

Because there is a small but measurable time-delay between the waves' being transmitted and received — it is about 3.3 µs per kilometer or 5.3 µs per mile — the scans at the receiver cannot be triggered by any local signal, such as the mains frequency. Synchronization has therefore to be carried out by the transmission of synchronizing pulses inserted into the video signals. A single *sync pulse* (as they are commonly abbreviated) is transmitted before every line scan starts, and a group of five longer pulses is transmitted before every field scan. These sync pulses trigger immediate flyback of the timebases, and so enable the scanning spot of the receiver CRT to be in the correct position when the video waveform arrives. The scans of transmitter and of receiver will thereafter remain in step provided the scanning waveforms remain undistorted. The complex sync-pulse sequence is shown in Figure 2.4. This period is often referred to as the *vertical blanking interval* (VBI).

A form of synchronization similar in principle, but somewhat different in detail, is required for colour signals.

34

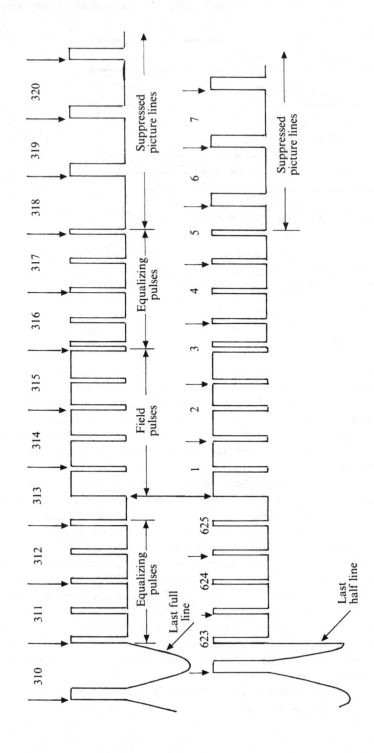

Figure 2.4   Sync pulses during vertical blanking interval

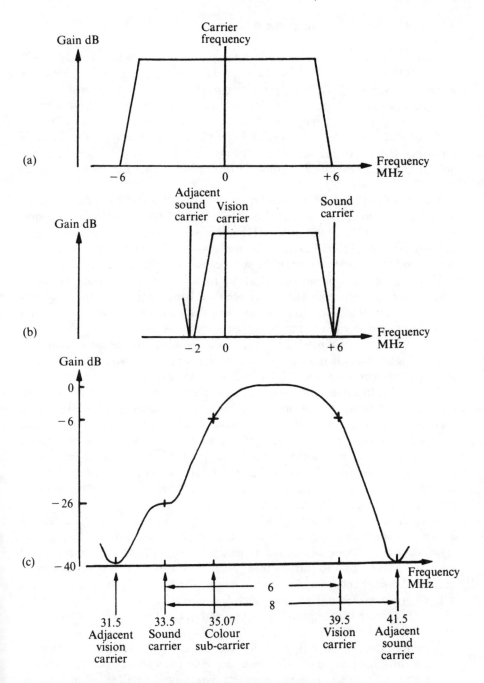

Figure 2.5   Vestigial sideband transmission and reception

Vestigial sideband transmission

TV transmissions in the UK are now all in the UHF channels of Band IV (470 to 582 MHz) and Band V (582 to 855 MHz), including the new Channel 5 service, the video signals being *amplitude-modulated*. As will be recalled from Volume 1, an unmodified amplitude-modulated transmission covering a bandwidth of 6 MHz would have also a set of sidebands extending to 6 MHz on either side of the carrier, as shown in Figure 2.5(a). Since only one set of these sidebands is used in the receiver, a technique called *vestigial sideband transmission* has been developed to conserve transmitter power and frequency spectrum.

The use of the vestigial sideband transmission technique alters the transmitted frequency spectrum to that shown in Figure 2.5(b). The lower sideband extends by 1.75 MHz instead of 6 MHz and provides a small guard band for the next lower channel sound carrier. Greater savings still would be possible if the carrier and lower sideband could be suppressed completely, but technical difficulties both in the transmitter and at the receiver become considerable when that is done. A compromise had to be achieved between the ideal and what was technically possible when the system was designed, and vestigial sideband transmission has proved in practice to be a good one.

The separate carrier needed for the *sound signal* is radiated at a frequency 6 MHz higher than that of the vision carrier, and is *frequency-modulated* to allow the intercarrier method to be used in the IF stages of the receiver.

The bandwidth of the complete transmitted signal, from the sound carrier to the outermost vestiges of the lower sideband, is now about 8 MHz. This figure therefore represents the *minimum possible* spacing which must be allowed between channels allotted to different transmitters. To avoid *adjacent channel interference* (adjacent channel is the name given to the two channels whose carriers are respectively 8 MHz higher and 8 MHz lower than the carrier under consideration), closely adjacent channels are always allotted to transmitters which are geographically well separated.

Nevertheless, interference can still arise from the fact that the sound carrier of a given transmission is spaced only 2 MHz from the vision carrier of an adjacent transmission. A strong adjacent-channel signal arising from this source would cause a 2 MHz signal to intrude into the video stages of a receiver, causing a pattern of disturbance to appear on the screen. The frequency modulation of the sound carrier makes this pattern much less obtrusive than it would be if the same carrier were still amplitude-modulated; but careful planning of frequency allocations is nevertheless required.

A complicating factor is that, in any given region of the UK, all the main TV transmissions have to be received on a single aerial, which has therefore to be cut to a length that will resonate over a comparatively small range of frequencies, ideally 88 MHz. All transmitters serving the given area can therefore (with a few exceptions) use only channels which lie within a total frequency spread of 88 MHz.

The IF stages of the receiver require a wide enough bandwidth to pass the whole 6 MHz bandwidth of the signal, from vision carrier to sound carrier. Their ideal response curve is therefore that shown in Figure 2.5(c). The standard IF centre-frequency in the UK is 36.5 MHz, with the vision carrier at 39.5 MHz and the sound carrier at 33.5 MHz. Figure 2.5(c) also shows the relative positions of the *colour sub-carrier* frequency at 35.07 MHz and the two adjacent channel carriers (31.5 and 41.5 MHz).

Note that the sound IF is always lower than the vision IF, even though the sound carrier itself is at a *higher* frequency than the vision carrier. This frequency inversion is caused by the fact that the local oscillator frequency is always, by convention, set higher than the frequency of the input signal. For example, if the vision carrier is at 550 MHz and the sound carrier at 556 MHz, the local oscillator (LO) frequency which would be used in the receiver would be 589.5 MHz. The vision IF then becomes 589.5 minus 550 = 39.5 MHz, and the sound IF, 589.5 − 556 = 33.5 MHz, as stated above.

Colour triangles and additive colour mixing

The basic nature of light and colour was introduced in Volume 1 of this series, to which the reader is referred for any necessary revision.

When light is analysed, it is found to have three components:

1 *brightness*, which is electrically equivalent to the amplitude of the light;
2 *hue*, the colour, which is determined by its wavelength; and
3 *saturation*, which is the variation in the depth of hue, that is, a pastel or deep shade.

A *saturated* colour is a deep shade, while a colour that is diluted with white light is said to be less saturated or *desaturated*. Most of the colours that occur in nature may be simulated either by adding different intensities of red, green and blue light, or by adding voltages that represent different levels of intensity of these three primary colours. This effect can be explained graphically using the so-called *colour triangle*, shown in Figure 2.6. The colours at the apexes of the triangle are saturated ones and can be considered as the colour centres of three beams of light. Along each edge of the triangle the colours merge and gradually change from blue to green to red to blue. The colours halfway along each edge are the secondary or complementary colours, cyan, yellow, and magenta, which are formed in the following way:

37

cyan = blue + green,
yellow = green + red,
magenta = red + blue.

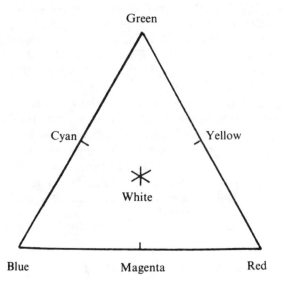

Figure 2.6   The colour triangle

These are described as complementary for the following reason: if two colours connected by a line drawn across the triangle and through its centre, are added, the resultant is white. Following any line from an edge towards the centre generates desaturated colours. The centre of the triangle is white = red + green + blue. Thus white is the ultimate desaturation of any colour displayed in the triangle.

To describe a given colour it is necessary to know its brightness or *luminance*, hue and saturation. Since luminance information is already transmitted in the monochrome TV system, it is only necessary to transmit further information on the hue and saturation to provide colour TV transmissions.

If three beams of red, green and blue light are added (as shown in Volume 1), the white only occurs in one small area. This shows that it requires certain intensities of the primary colours to generate white light. In the electrical sense, if each of the three primaries is generated by 1 V, then white light will result from the addition of 0.3 V red, 0.59 V green and 0.11 V blue to produce the equivalent of 1 V of white. This can be expressed algebraically as:

$$E'_Y = 0.3E'_R + 0.59E'_G + 0.11E'_B.$$

where E' represents the *gamma-corrected* voltage, Y signifies the luminance component and the suffixes R, G, and B refer to the red, green and blue components respectively.

Exercise 2.1

Connect a pattern generator to a colour TV receiver and display the standard colour bar pattern. Sketch and record the colour order of the display. Switch off the CRT guns, first in pairs and then singly, and again record the displayed patterns. Compare the results with the colour triangle of Figure 2.6.

Gamma correction

Ideally, the TV transmission and reception system should be linear. That is, the light output from the receiver CRT should be proportional to the light input to the camera tube. However, the receiver CRT has a beam current (and hence light output) to grid voltage ratio that is approximately a square law. To compensate for this, the transmitted video signal is pre-distorted in a complementary manner, described as *gamma correction*, to produce an overall linear system. To distinguish between the raw uncorrected camera voltages and the gamma-corrected ones, it is a convention to use primes ('). Thus E_y is an uncorrected luminance voltage, whilst E_Y' is a gamma-corrected one.

Colour TV camera and colour difference signals

If additive colour mixing is to form the basis of a colour TV system, then information about the intensities of the three primary colours has to be transmitted along with the luminance signal. The transmission of four such components of the video signal would require the use of excessive bandwidth and complex transmission coders and receivers. The concept of *colour difference* signals was developed to overcome these problems.

It is possible to have a number of different camera arrangements, varying from four tubes (one for luminance and one for each primary colour) down to a single tube that requires a complex beam switching system to generate the three primary colours. Figure 2.7 shows one such arrangement and the way of generating the colour difference signals. This is a particularly useful method, as the three primaries can be regenerated from the luminance and two colour difference signals as follows, using suitable circuits or matrices.

$$(E_R' - E_Y') + E_Y' = E_R'$$
$$(E_B' - E_Y') + E_Y' = E_B', \text{ and since } E_Y' = E_R' + E_G' + E_B',$$
$$E_Y' - (E_R' + E_B') = E_G'.$$

Referring to Figure 2.7, the light from the scene to be televised is focused on to a mirror system. The first mirror is *half-silvered* to split the light beam into two parts. One part activates the luminance tube, whose output is amplified and

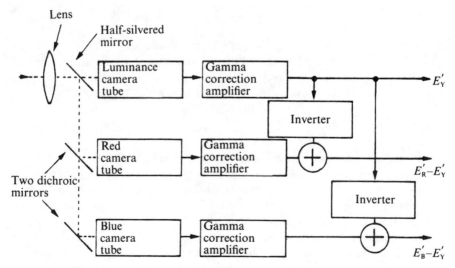

Figure 2.7   A camera arrangement

gamma-corrected to produce the E'_Y signal. Dichroic mirrors have a part-mirror, part-filter action: that is, they reflect only a narrow range of wavelengths and allow others to pass through. Here, two such mirrors are used to separate out the red and blue components. Each signal is then amplified and gamma-corrected. An inverted version of the luminance signal is added to both the red and blue signals to generate the two colour-difference signals $(E'_R - E'_Y)$ and $(E'_B - E'_Y)$.

Modulation by the colour-difference signals

The colour-difference signals are transmitted by a process known as *quadrature amplitude modulation* (QAM), *double sideband suppressed carrier* (DSSC). Simply stated, this means that each of the two colour difference signals modulates one of two versions of the same carrier frequency, which differ in phase by a quadrant (90°).

The modulation process is carried out with circuits known as *balanced modulators* whose action is shown in Figure 2.8. Normal amplitude modulation produces carrier and upper and lower side frequencies (Figure 2.8(b)), but the balanced modulator cancels out the carrier component. The products of modulation are now added together to produce the composite known as the *chrominance signal*. This phasor addition is shown in Figure 2.9(a). The amplitude of the chrominance signal represents the degree of colour saturation, while the phase angle represents its actual colour. The colour coding method used in the UK and most of Europe is the PAL (Phase Alternation Line by line) system.

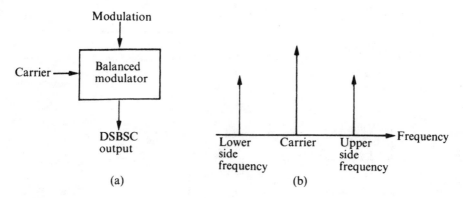

Figure 2.8   Action of balanced modulator

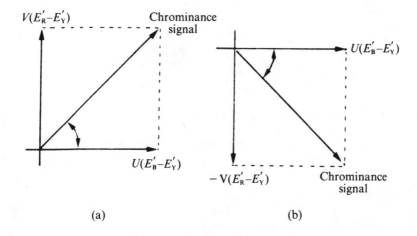

Figure 2.9   The PAL system chrominance signal: (a) NTSC line; (b) PAL line

This was developed from the American NTSC (National Television Standards Committee) system. In the PAL system, the V component (derived from $E_R' - E_Y'$) of the chrominance signal is inverted or phase-shifted by 180° on alternate lines. Figure 2.9 shows the phasor diagrams for the same colour on successive lines. This system was developed to minimize the effects of amplitude and phase errors of the received chrominance signal, which in the NTSC system produces significant colour errors. By comparison, the French SECAM system uses frequency modulation of two separate subcarriers, one for each colour difference signal. These are then transmitted individually and line sequentially.

Figure 2.10 shows the basic coding system for the PAL signal. The two colour-difference signals modulate two versions of the 4.433 618 75 MHz sub-

41

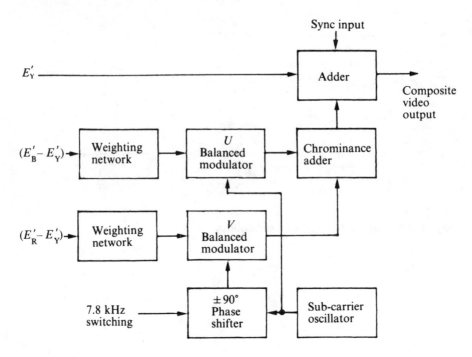

Figure 2.10   The basic PAL system coder

carrier via the balanced modulators to produce DSSC outputs, the sub-carrier signal to the V modulator being switched $\pm$ 90° line-by-line to produce the necessary signal inversion. The switching at approximately 7.8 kHz is derived by dividing by two from the line timebase generator. The modulation products are then added to produce the chrominance signal. This latter is then added to the luminance signal together with the line and field sync pulses to form the composite video signal. The combined amplitudes of the luminance and chrominance signals may now be great enough to produce over-modulation. To avoid this, both colour-difference signals are scaled down in amplitude by fixed ratios in the weighting networks, only to be expanded by a complementary ratio in the receiver. The weighted signals are $U = 0.493(E_B' - E_Y')$ and $V = 0.887(E_R' - E_Y')$ respectively. The complementary receiver deweighting ratios are thus the inverse of these numbers, 2.028 and 1.127 respectively.

Colour burst

To enable the colour TV (CTV) receiver to decode the PAL system line inversion sequence correctly, a further synchronizing signal is added to the coder output shown in Figure 2.10. Ten cycles of the colour sub-carrier frequency at 4.433 618 75 MHz are added during the sync pulse back-porch period shown in

Figure 2.11 and 2.13. This burst, which is switched in phase between 135° and 225° relative to the $+U$ axis, line by line, is called the *swinging burst*, the phase inverted lines being identified by the 225° burst.

Compatibility

Whilst the CTV system was being developed there were several millions of monochrome receivers in use throughout the world. To introduce colour in a non-compatible manner would have been economically unacceptable. To make the systems compatible, the CTV signal must produce a black-and-white picture on a monochrome set. In the reverse sense, there are still a lot of old black-and-white films suitable for transmission. Therefore, to be *reverse-compatible*, the CTV receiver must produce a black-and-white picture from a monochrome signal.

The grey scale

As colour TV developed, it became increasingly necessary to provide some form of standard test signal for system and receiver setting up, and for comparison purposes. One of the most useful is the *colour bars* pattern which consists of a series of vertical strips of the primary and second colours with black and white. These are arranged in a pattern of reducing luminance from left to right. On the monochrome receiver these will be displayed as bars of varying shades of grey. Figure 2.11 shows one line of the video signal without chrominance information and showing the grey scale for 100% saturated, 100% amplitude, colour bars. The steps of the signal can be calculated from the basic luminance equation:

$$E'_Y = 0.3E'_R + 0.59E'_G + 0.11E'_B$$

as shown in Table 2.1. Because of the 100%/100% relationship, the output for each primary colour is at a maximum of unity.

Exercise 2.2

Examine the colour-bar test pattern on colour and monochrome receivers and an oscilloscope. Relate the video waveform to what is displayed on each receiver. Find out what use can be made of the grey scale pattern for setting up a colour TV receiver.

Make a summary list of the UK PAL system standards for both colour and monochrome transmissions.

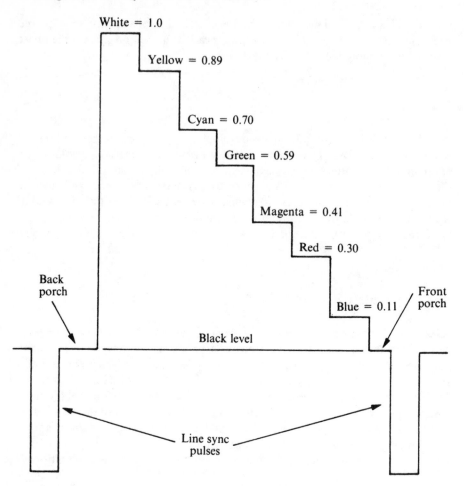

Figure 2.11   The grey scale

Table 2.1   The grey scale

Colour bar	E'_R	E'_G	E'_B	$E'_Y = 0.3E'_R + 0.59E'_G + 0.11E'_B$	
White	1	1	1	0.3 + 0.59 + 0.11	= 1
Yellow	1	1	0	0.3 + 0.59 + 0	= 0.89
Cyan	0	1	1	0 + 0.59 + 0.11	= 0.70
Green	0	1	0	0 + 0.59 + 0	= 0.59
Magenta	1	0	1	0.3 + 0 + 0.11	= 0.41
Red	1	0	0	0.3 + 0 + 0	= 0.30
Blue	0	0	1	0 + 0 + 0.11	= 0.11
Black	0	0	0	0 + 0 + 0	= 0

The composite video signal

The chrominance signal is formed by the phasor addition of the 90° out-of-phase U and V components from $\sqrt{(U^2 + V^2)}$. Figure 2.12 shows how this is achieved for 100% saturated colour bars. This signal is then superimposed upon the luminance component of Figure 2.11, to produce the composite signal shown in Figure 2.13. In addition, this diagram also shows the position of the swinging burst on the sync-pulse back porch.

Exercise 2.3

The reader should practise drawing and calculating the waveform shown in Figure 2.11, because this is a popular examination question.

Test questions

1 For the UK UHF television systems state:
   (a) the chrominance subcarrier frequency;
   (b) the two functions performed by the colour burst;
   (c) the number of cycles transmitted during the burst period.

2 Draw a block diagram and explain how the two colour-difference signals $(R-Y)$ and $(B-Y)$ are obtained from the R, G, and B camera outputs.

3 (a) State why the colour-difference signals are weighted to obtain the U and V signal components.
   (b) How does the use of DSSC modulation of the chrominance signal aid the transmission of a monochrome signal from a colour-television transmitter?

4 Sketch the three time-related waveforms that would be required to produce the video line XX in Figure 2.14. Identify each waveform, label the peak amplitudes and indicate the periodic time.

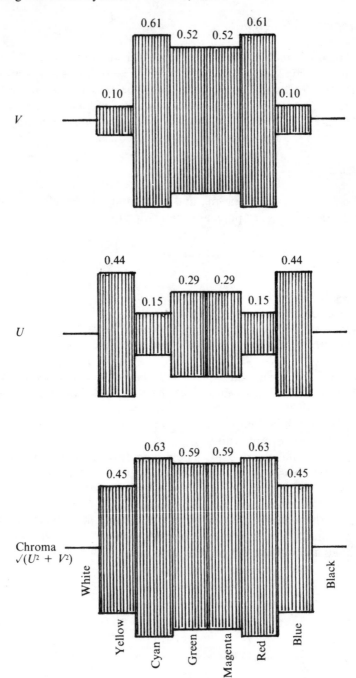

Figure 2.12   Forming the chroma signal

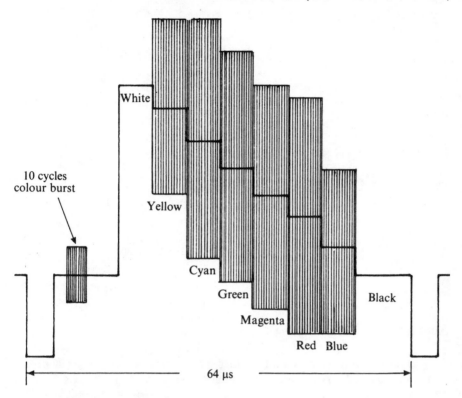

Figure 2.13 Composite video signal

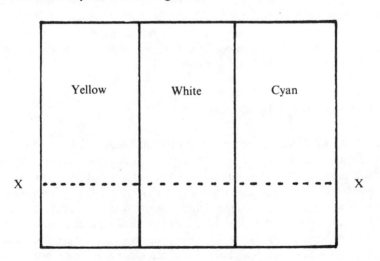

Figure 2.14 Screen display for test question 4

3 Television circuits

Summary

Demodulation and demodulators. Video and luminance amplifiers. Chrominance amplifiers. Luminance/chrominance signal processing with ICs.

Much of the signal processing covered in this chapter is commonly embedded within Application-Specific Integrated Circuits (ASICs) with a very high level of integration. In spite of this, many of the discrete-component circuits from the previous series have been retained because, with these, it is much easier to explain the fundamental principles involved.

Decoding the composite signals

Figure 3.1 shows a block diagram that represents the signal processing covered in this chapter. The IF signal input carries the sound, luminance, colour and sync information within its modulation. The locations of these various components within the signal spectrum were shown in Figure 2.5. For both the colour and the monochrome receiver, the demodulator stage will separate the sound and video components, while the video amplifier will generally act as an amplifier and level detector, the latter function producing the sync pulse train. The processing in the chrominance channel is much more complex, since the colour information is coded as colour difference signals carried by a form of suppressed carrier modulation.

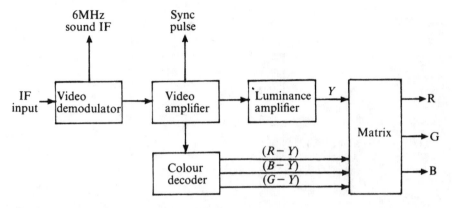

Figure 3.1 Composite-signal decoding stages

Demodulation and demodulators

The purpose of the demodulator is to recover the information from the modulated wave. In colour TV, this signal carries the luminance (luma), chrominance (chroma) and sound information. All three need to be recovered and separated from each other with the minimum of mutual interference. The use of different forms of modulation allows this to be achieved in a fairly straightforward way. Because the vision signal is amplitude-modulated, the luminance signal can be recovered with a simple diode detector. After filtering, the demodulated (baseband) signal will contain luminance information, together with the chrominance signal still quadrature-amplitude-modulated on its subcarrier. There will also be a 6 MHz beat frequency between the two main carriers which is frequency-modulated with the sound information (the *inter-carrier* sound signal). Although the diode detector was the norm for many years, the higher quality standards demanded by modern display tubes mean that the diode no longer performs adequately, and so it has given way to the *synchronous demodulator*.

Negative modulation is used for the vision signal in the UK TV system. This means that sync pulses are represented by maximum amplitude and peak white by the minimum. This is shown in Figure 3.2(a). Note that peak white represents a level of 20% relative to sync-pulse tips. This amplitude allowance provides space for the chrominance signal to be added, without driving the carrier amplitude down to zero on signal peaks. The polarity of the video signal after demodulation depends on the part of the envelope that the detector follows. Either polarity may be required depending on the number of inverting signal amplifiers that are needed between the detector and the CRT. Using the diode detector as a simple example, Figure 3.2(b) shows the recovered baseband signal.

The chrominance signal is extracted from the baseband with a *notch filter*. This is so-called because of the shape of the luminance response curve due to its

49

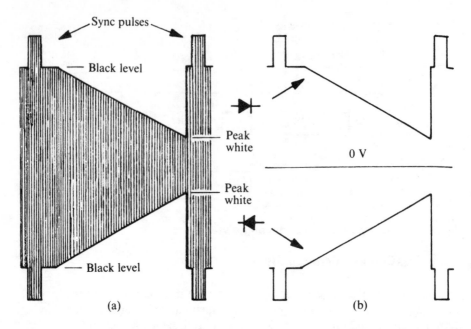

Figure 3.2 (a) Negative-modulated vision signal; (b) demodulated video signal

action. This action is needed to minimize the mutual interference between the luminance and chrominance signals. The baseband frequency response is shown in Figure 3.3 together with the 6 MHz inter-carrier sound FM signal. The sound carrier is now amplified and limited to remove any amplitude modulation

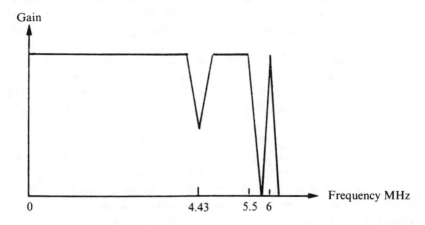

Figure 3.3 Demodulated baseband signal spectrum showing 'notch' in luma signal

caused by the vision signal. Finally the sound is demodulated in the conventional FM manner described in Chapter 1.

Figure 3.4 shows the principle used to separate out the three components of the signal. L_1C_1 is a series-resonant, low-impedance path for the 6 MHz sound,

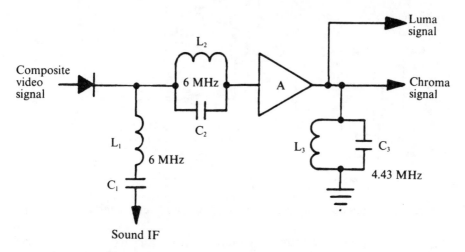

Figure 3.4   Principle of sound and chroma signals take-off

while L_2C_2 forms a high-impedance block, preventing the FM signal from entering either the luminance or chrominance channels. In a typical discrete-circuit CTV, a buffer amplifier (emitter follower) will be used to prevent the video-signal circuits from loading the detector stage. L_3C_3 forms a resonant, high-impedance circuit to select out the QAM chrominance component. The signal is then low-pass filtered to provide the luminance component.

Figure 3.5 shows how the technique was incorporated into a practical amplifier circuit. The demodulated signal is applied to the base of a transistor which has a tuned load of 6 MHz, and is thus fed to the sound channel. From the buffered video signal appearing at the emitter of the same transistor, the 6 MHz intercarrier is filtered off from the luma and chroma signal by the series-resonant circuit L_2-C_2.

Figure 3.6 shows how the integrated circuit, TDA 3540, is organized to deal with the modulated IF signal. The first stage, a gain-controlled IF amplifier, feeds a synchronous demodulator, where the signal is effectively multiplied by its own carrier frequency, controlled by L_2, C_{23}. This produces an output containing the composite video signal, plus the 6 MHz FM sound intercarrier, with relatively little distortion. The output signal components are separated in the manner shown in Figure 3.4, the sound take-off point being at L_4. The AFC section contains its own synchronous demodulator with its frequency controlled by L_3,C_{25}. An AFC on/off switch (pin 6) is provided so that this system can be disabled during initial channel tuning. The on-chip AGC system not only controls the gain of the IC IF amplifier

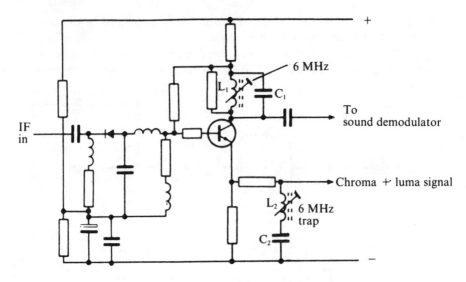

Figure 3.5   Practical example of the separation of sound and vision signals

stages, but also provides an output to control the gain of the RF stages in the tuner unit. Impulsive-type interference shows up as white spots on the CRT. These are detected within the chip and inverted to black spots that are less objectionable.

Video luminance amplification

The luminance bandwidth for most colour TV receivers now covers d.c. to 5.5 MHz, but generally with a notch in the frequency response at 4.433 MHz. This feature, shown in Figure 3.3, is due to the filter action of the colour subcarrier circuits.

The need for such a wide bandwidth was explained in Chapter 2. The highest frequencies are needed to provide fine picture detail up to a maximum of 383.5 complete cycles per line. Response right down to d.c. is needed because the overall brightness of the picture is represented by the *average* d.c. level of the signal as a whole. An incorrect d.c. level will thus produce unsatisfactory pictures, with no true blacks present in scenes intended to represent night-time.

Monochrome discrete-component video amplifiers comprise, in most circuit designs, two transistors. A wide bandwidth is achieved by using resistors of low value as collector loads, with no tuned circuits apart from the filter traps. The reason why load resistors must have low values is because otherwise the stray capacitances which are always present in an amplifier would by-pass the signal. A 5 pF stray capacitance, for example, will have a reactance of only 5k8 at 5.5 MHz; but such a low value of stray capacitance would only be found if the

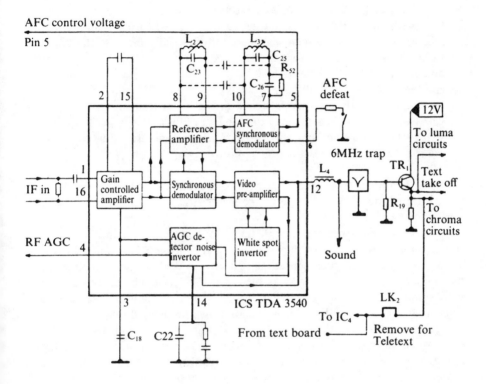

Figure 3.6   The demodulation stage

video-amplifier stage were not coupled to any other circuits. In practice, the signal will be fed from the video amplifier to the cathode of the tube, so that a considerable amount of further stray capacitance is added.

When the load resistance equals the reactance of the stray capacitance, the response of the receiver is lowered by 3dB; and to be only 3dB down at 5.5 MHz is something of an achievement in a mass-produced receiver.

The value of the collector load resistors is therefore critical. It must be kept lower than the reactance of any stray capacitances likely to be encountered in the stage if the frequency response of the receiver as a whole is not to be badly impaired.

53

Techniques used to extend bandwidth include the following:

1 very low resistance values for all collector loads;
2 the use of small inductors called *peaking chokes* which have an increasing reactance at higher frequencies (these automatically increase the effective value of the load, to combat the gain-reducing effects of the stray capacitances);
3 emitter resistors decoupled by a small capacitor. This enables the emitter resistor to provide negative feedback at the lower frequencies, so keeping gain low. But as the reactance of the capacitor falls at the higher frequencies, so it cuts down the amount of negative feedback generated, thus compensating for the frequency losses at the collector.

Figure 3.7 illustrates the video amplifier of a monochrome portable. The first stage is an emitter-follower, with the collector circuit used to effect sync-pulse separation (see later).

The video output stage uses a single transistor connected to a 95 V supply. The collector load resistor has a value of 6k8, and the 'no-signal' standing current is about 7 mA. Frequency compensation is effected by the low-value capacitor C_1 in the emitter circuit, and VR_1 acts as the *contrast* control. The blanking pulse is used to shut down the video amplifier during the vertical blanking interval (VBI). This avoids field-flyback lines being superimposed on the picture.

Figure 3.8 illustrates the *luminance amplifier* stage of a colour receiver. Each of the three cathodes of the colour CRT is driven by a signal which is composed, in the correct proportions, of the luminance signal and of one of the three colour signals. The luminance amplifier cannot therefore drive the cathode of a CRT directly, but is coupled instead to a signal-adding *matrix* stage.

The luminance amplifier of Figure 3.8 uses a single voltage amplifier Tr_1 with a 1k load, and the transistor Tr_2 connected as an emitter-follower, to drive the luminance amplifier Tr_3. Tr_3 collector load R_5 is a 1k8 resistor which in turn is coupled by a delay line to the matrix stage. L_2 and L_4 form a peaking choke circuit designed to combat the effects of stray capacitance.

Note that the coupling through C_2 results in the d.c. level of the signal being lost. It is restored by the diode D_1, the d.c. level of the cathode of this diode being adjustable by manipulation of the *brightness* control.

Emitter-followers which have good signal-isolating properties are used as buffer amplifiers in the video stages. Their high input impedance places very little load on the demodulator circuit, and their low output impedance enables them to drive circuits with relatively large stray capacitances.

The video output transistor (Tr_3 in Figure 3.8) requires a higher-voltage supply than do previous stages, to enable it to drive the grid of the CRT fully between black level (which is cut-off, at maybe $+80$ V) and peak-white (at around $+10$ V). The transistor used in this stage must therefore be of a type

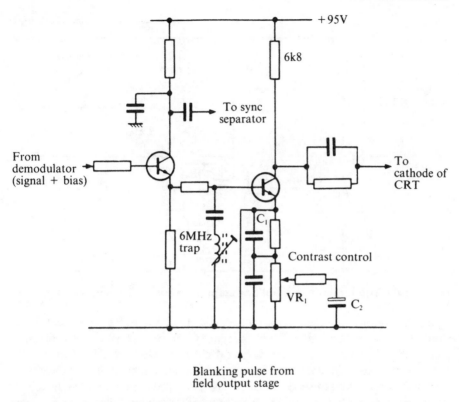

Figure 3.7 A video amplifier stage

capable of handling the levels of power and voltage which will be encountered, and of giving the frequency response required.

Cross-colour, cross-luminance effects

If any high-frequency luminance signal reaches the chrominance channel, it will be processed as chroma information and be displayed as false colours. This can usually be seen on the 4.5 MHz luminance bars of the test card. Similarly, if any chroma signal reaches the luma channel it will be processed as a high-frequency luma component. This effect is displayed as moiré patterning, particularly on fine picture detail.

Chrominance amplification

The stringent demands of gain with wide bandwidth do not apply to chrominance amplification. The human eye is so constructed that it is more sensitive to

55

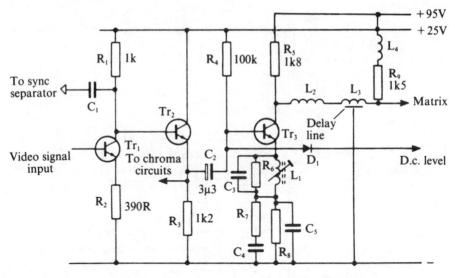

Figure 3.8   The luminance amplifier stage of a colour TV receiver

light-intensity variation than changes of actual colour. Subjective experiments have shown that satisfactory colour TV images can be transmitted using a chrominance bandwidth of about ± 1 MHz. A typical chrominance amplifier of a discrete component stage might contain three common emitter amplifiers. Tuned circuits, if used, will be resonant at 4.433 MHz and heavily damped to provide a bandwidth of about ± 1 MHz, often a 6 MHz sound trap will also be included. Alternatively, resistance capacity coupling may be used together with some form of HF peaking circuit. The general principle of the recovery of the two weighted colour-difference components is explained with the aid of Figure 3.9. The composite chroma signal is

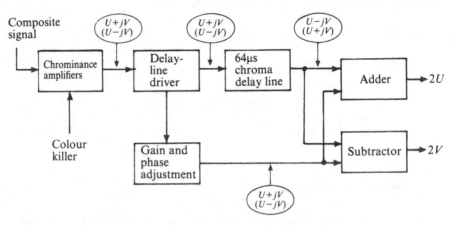

Figure 3.9   Decoding the colour-difference components

split into two paths, one through a 64 µs ultrasonic delay line (one horizontal line period) and the other through a gain- and phase-balancing circuit. This ensures that both paths to the adder and subtractor have the same characteristics. It will be recalled that, in the PAL system, the V signal is inverted on alternate lines, and this is indicated by the signals $U+jV$ and $(U-jV)$ on successive lines ($\pm j$ signifies the $\pm$ 90° phase switching). While the $U+jV$ component is entering the delay line, the previous line signal $U-jV$ is emerging. The adder stage is thus presented with two line signals simultaneously: $(U-jV)$ from the delay line, and $U+jV$ from the direct path. This output is therefore equal to $2U$. At the same time, the output from the subtractor is the difference, equal to $2jV$. It is left to the reader to confirm that the same arithmetic operations on the two signals $U-jV$ and $U+jV$ yield the same results.

The U and V components are now synchronously demodulated and de-weighted to produce the $B' - Y'$ and $R' - Y'$ colour-difference signals. The $G' - Y'$ colour-difference component is recovered by a matrix circuit that performs the following arithmetic operation using attenuation, inversion and adding:

$$1(Y)' = 0.30R' + 0.59G' + 0.11B'$$
$$0.30\,Y' + 0.59\,Y' + 0.11\,Y' = 0.30R' + 0.59G' + 0.11B'$$
$$0.30(R' - Y') + 0.59(G' - Y') + 0.11(B' - Y') = 0$$
$$0.59(G' - Y') = -0.30(R' - Y') - 0.11(B' - Y')$$
$$\therefore (G' - Y') = \frac{-0.30}{0.59}(R' - Y') - \frac{0.11}{0.59}(B' - Y')$$
$$= -0.51(R' - Y') - 0.186(B' - Y')$$

Given that the luminance and three colour-difference signals are now available, the individual R, G and B components can be recovered using a further matrix in the following manner:

$$(R' - Y') + Y' = R', \quad (G' - Y') + Y' = G', \quad (B' - Y') = B'.$$

The *colour killer* is a circuit that detects the colour-burst sync component to produce a d.c. bias. This is applied to one of the chroma amplifiers to enable the chroma signal to reach the colour decoder circuits. In the absence of a colour burst (on a monochrome picture), the chroma circuits are thus shut down. This prevents colour noise from producing an interference pattern on the luminance signal.

Luminance delay line

The difference in frequency response of the luminance and chrominance ampli-fiers causes the signal to pass through the luminance amplifier faster than through the colour decoder. If the two components arrived at the CRT in this fashion, the mis-registration would produce a *ghosting* effect. Compensation is applied in the form of an artificial transmission line that delays the luminance component by about 0.7 µs.

The integrated circuit (TDA 3562A) shown in Figure 3.10 represents the high degree of integration achieved for the modern colour TV receiver. Basically, the inputs are luminance and chrominance signals, and the outputs are the red, green and blue colour signals. The chrominance signal is demodulated to recover the U and V components, which are then *de-weighted*. The two colour-difference signals (($B - Y$) and ($R - Y$)) are then inverted and added in the correct proportions to produce the missing ($G - Y$) signal in the ($G - Y$) matrix. The luminance signal from the demodulator has been delayed by about 0.7 µs before being input to pin 8. The luminance signal and the three colour-difference signals are then added in the second matrix (RGB matrix) to provide the RGB signals needed to drive the d.c.-coupled amplifiers between this IC and the three cathodes of the CRT. On-chip contrast, brightness and colour controls are provided for, these being activated by d.c. potentials from the user-variable controls. Provision is also made for RGB inputs from a teletext decoder, a video camera, or a computer, when the receiver can be used as a *colour monitor*.

Multi-standard signal processing

A key component of multi-standard signal processing is shown in Figures 3.11 and 3.12. The use of this IC allows the production of a universal TV receiver to be based on a standard chassis. The selection of the appropriate television standard is automatic and based on a sequential recognition algorithm. However, for low signal areas where this technique might fail, provision is made for manual selection.

The composite video signal is applied via switchable filters to the input at pin 15, the filters separating out the luma and chroma components according to the selected standard. The decoder scans the input signal sequentially, first testing for the frequency-modulated subcarriers of a SECAM signal at 3.9 and 4.756 MHz. Both the NTSC and PAL systems use quadrature-amplitude modulation (QAM) but can be separated by detecting the subcarriers of 3.579 45 MHz (NTSC) or 4.433 618 75 MHz (PAL). However, since non-standard NTSC, which uses the same PAL subcarrier frequency, may be encountered, the alternate line-phase reversal of the V chroma signal component can be used to detect the PAL standard.

The remainder of the on-chip processing then follows standard practice to produce colour-difference signals at the output. In spite of this high level of integration, it can be seen that many external components are still necessary. These include chrominance delay line for PAL decoding, tuned circuits and crystals to control the various subcarrier oscillators, and standards filters. All of

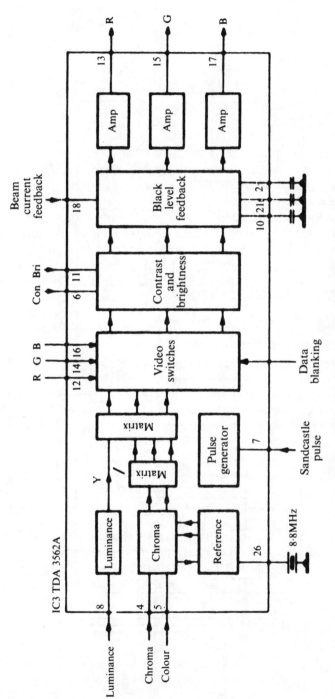

Figure 3.10 Luminance and chrominance processing

59

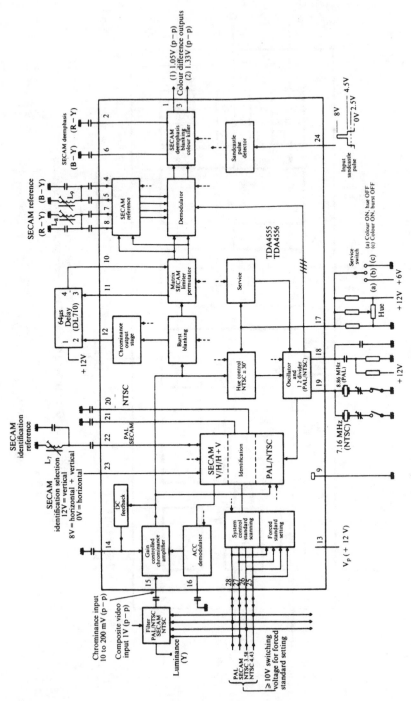

Figure 3.11 Multi-standard processor (courtesy of Philips Components Ltd)

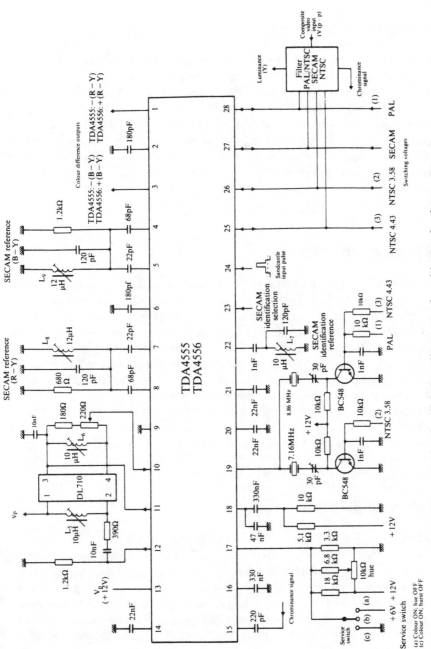

Figure 3.12 Application diagram of multi-standard processor

61

these require careful alignment if a high standard of colour processing is to be achieved.

Small-signal combination IC

The IC shown in Figures 3.13 and 3.14 represents an alternative approach to circuit integration. Using this chip, a TV receiver can be produced by the addition of little more than a tuner, a colour decoder, video, audio and timebase output stages and a CRT. The chip requires inputs of sound and vision at i.f. and provides outputs of composite video and audio, synchronized field and line timebase drives, as well as providing for automatic gain and frequency control of the tuner unit. The circuit even automatically selects for 50 or 60 Hz field timebases. Although a high level of integration has been achieved, there is still a significant number of external components. Some of these require accurate alignment and can also give rise to fault conditions.

CRT colour drive amplifiers

In this respect there are two methods that can be used. The first is *RGB drive*, in which the three signals *R, G* and *B* are applied separately to the three cathodes, while the three grids are strapped in parallel. The alternative is *colour-difference drive*, where the luminance signal (*Y*) is applied to common cathodes, and the three colour-difference signals $(R-Y)$, $(G-Y)$ and $(B-Y)$ are applied to the three separate grids. The latter method has the advantage that the R,G,B matrix is no longer required, as this is provided by the grid-to-cathode circuit. Also, on monochrome signals, the colour-difference signals are zero so that only the luminance signal is present. However, because this method requires about twice as much drive voltage as for RGB drive, the RGB method is now almost universal. Note also that when grid/cathode drive is used, the matrixing is effectively being performed by the grid/cathode action, and this will be less linear than the action of a resistive matrix.

Single-stage RGB drive circuit

The circuit shown in Figure 3.15 is suitable for a small portable colour TV receiver. It uses a common emitter amplifier but with a small value of bypass capacitor to ensure gain stability from d.c. up to about 5 MHz. Three identical stages, one for each colour, are used between the colour-processing IC and the CRT cathodes. The three preset background controls are used to produce a white raster with no input signal. Spark gaps are incorporated in the circuit to protect the transistors from damage due to *flash over* discharges within the CRT.

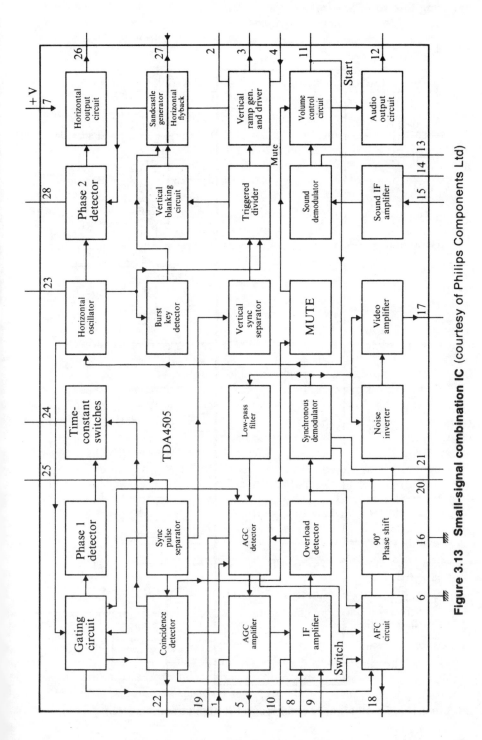

Figure 3.13 Small-signal combination IC (courtesy of Philips Components Ltd)

63

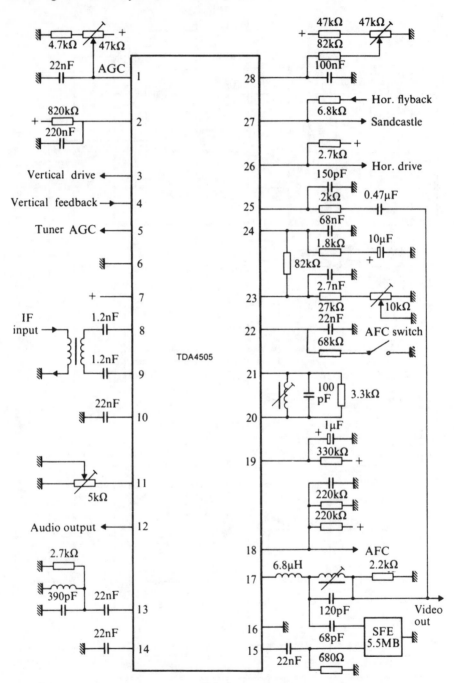

Figure 3.14   Small-signal combination IC: application diagram

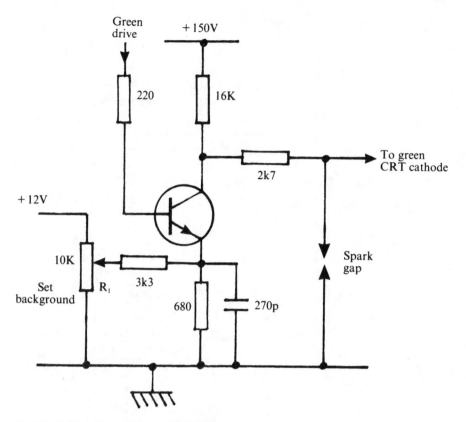

Figure 3.15   Single-stage RGB drive circuit

Multi-stage RGB drive circuits

The circuit shown in Figure 3.16(a) is designed for the large-screen, wide-angle CRT. It uses three identical stages coupled between the colour-processing IC and the CRT cathodes. In such applications, it is necessary to produce a signal swing between black level and peak white of about 100 V. The necessary gain is achieved by using a compound-pair amplifier that consists of a common emitter amplifier (Tr_1) with a high value of collector load (R_2) and direct coupling to the emitter-follower stage (Tr_2). This provides a low-impedance drive circuit for the colour signal which is applied to the CRT cathode via the diode D_1. The circuit is gain-stabilized by the resistor ratio $R_3:R_1$. The gains of the three stages are balanced by the preset resistors (R_1) to obtain a white raster with no input signal. Tr_3 forms part of the beam current protection circuit (see Chapter 5).

The circuit depicted in Figure 3.16(b) represents one of three identical stages, of which only the blue amplifier is shown for simplicity. Each obtains the necessary gain and bandwidth by using a common-emitter, common-base com-

65

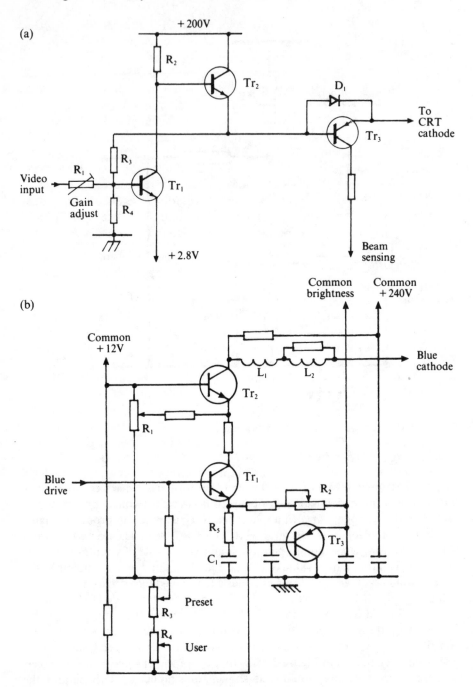

Figure 3.16 Multi-stage RGB drive amplifiers

pound-pair amplifier, with peaking chokes (L_1 and L_2) and negative feedback via R_5 and C_1. Common brightness control is effected via the d.c.-coupled amplifier Tr_3 and the series arrangement of preset and user controls. Each amplifier stage contains two preset controls R_1 and R_2. These would be adjusted according to manufacturers' instructions using meters to set the CRT beam cut-off points and the peak beam current. Effectively the aim is to produce accurate grey-scale tracking using a suitable test pattern. R_1 is used to set the point at which each gun in turn just starts to conduct: i.e., it sets the overall black level. The gain control R_2 is then used to set the peak white level.

Finally, the brightness and contrast controls are adjusted for equal brightness changes between steps. The dark or low-light areas are observed for signs of coloration and the R_1 (A_1) controls adjusted accordingly. Similarly, the bright or highlight areas are observed and the R_2 (gain) controls adjusted as necessary. This should leave an equal-gradation grey-scale pattern without false colouring.

Universal interface

This device, known variously as the SCART or Peritelevision socket or Euro-connector provides a standard 21-pin interface to allow the interconnection of several peripheral devices to the TV system. A limited degree of flexibility is provided for, but the status of most pins is standardized (see Figure 3.17). The concept allows for interconnections to be made at RGB and baseband level so that no remodulators are needed to provide input to the UHF aerial socket. The use of these tends to degrade the picture quality due to mixer noise and tuner drift.

Video-switching matrix

A number of semiconductor manufacturers have produced signal-routeing matrix chips that are controlled through the normal receiver remote-control system. These allow a variety of peripheral devices to be permanently wired in a tidy manner. The general arrangement for this technique is shown in Figure 3.18, where a remote-control input is decoded and the necessary cross-points selected to connect any input to any output. The switchers typically have a bandwidth of about 10 MHz and a cross-talk isolation between unused inputs and outputs that is in the order of -55 dB.

User controls

The position of the three variable resistors that form the contrast, brightness and colour controls varies considerably from model to model. The saturation or colour control is invariably in a position so that it adjusts the gain of the

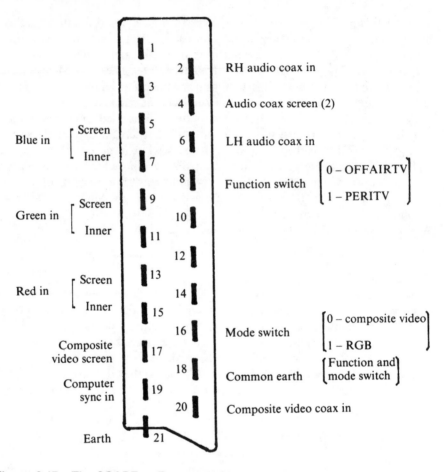

Figure 3.17   The SCART or Euroconnector

chrominance amplifier. The contrast control is usually placed so that it varies the gain of the video or luminance amplifiers. The basic requirement of the brightness control is that it should vary the grid-to-cathode voltage of the CRT. In a monochrome receiver its position is thus obvious. However, in the colour TV receiver, the adjustment of the grid-to-cathode voltage has to be made simultaneously to the three guns. This adjustment is thus usually made indirectly through d.c.-coupled amplifiers.

Fault finding in the signal processing stages

The service engineer should be familiar with the waveforms to be expected at different points in a working receiver. The study and sketching of these can

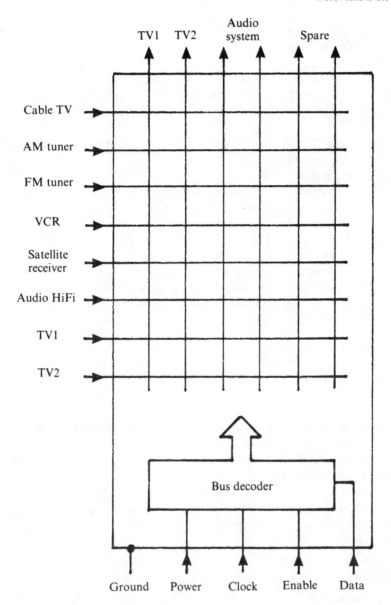

Figure 3.18   Video-switching matrix

provide a valuable reference for future fault-finding. Personal copies of system circuit diagrams can usefully be annotated with comments as one gathers experience. This then provides a valuable future reference source. Tracing faults

69

in a highly integrated system is difficult unless great care is taken. The following are some useful guidelines for dealing with receivers using ICs.

1   Identify the faulty area from the symptoms observed.
2   Check all d.c. voltages around the suspect IC. These devices are commonly d.c.-coupled throughout, so that incorrect d.c. levels will affect the signal paths.
3   Carefully check signal waveforms at input and output, and compare them with those given in the service data.
4   Check the circuitry associated with the input/output of the suspect chip. If necessary, disconnect the feeds and note the effect on the d.c. readings.
5   Take care not to introduce short circuits with the test prods.
6   Avoid overheating the device with a soldering iron.
7   After removing the suspect IC, check the circuit board for short circuits and correct d.c. supply level and signal inputs.
8   Unless the IC is socket-mounted, removal by desoldering will increase the heat stresses so that even if the device is currently serviceable, it becomes susceptible to premature failure. It is thus usually false economy to replace such displaced components. This is particularly true for surface-mounted devices.

Exercise 3.1

(a)   Using an oscilloscope, examine and record the waveforms found at the various test points in a working colour TV receiver.
(b)   Study the effects on the waveforms as the brightness, contrast and colour controls are adjusted.
(c)   Study the effects on the waveforms and the CRT display as the various presets around the colour decoder IC are mis-adjusted.
(d)   Carry out the grey-scale and drive adjustments on a working colour TV receiver.

Test questions

1   The PAL colour TV receiver conventionally has two delay lines.
    (a)   State the part of the circuit in which they are used.
    (b)   State the delay period for each.
    (c)   What would be the visible effect if each were separately open circuit?
    (d)   Briefly explain why each is used in the receiver.

2 For a PAL TV receiver:
   (a) state the colour subcarrier frequency;
   (b) state the typical bandwidth of the chrominance amplifiers;
   (c) state the typical bandwidth of the luminance amplifiers;
   (d) state the bandwidth of the RED amplifier in an RGB drive circuit.

3 State the function of:
   (a) the 4.433 MHz filter in the luminance amplifier stage;
   (b) the 6 MHz filter in the luminance circuit.
   (c) What would be the effect if high-frequency information from the luminance signal were to reach the chrominance amplifiers?

4 (a) Briefly describe the operation of the two matrix circuits used in the colour-processing stages of a PAL receiver.
   (b) Why are blanking pulses applied to the RGB matrix circuit?

4 Timebases and synchronizing circuits

Summary

Sync pulses. Separation. Circuits. Flywheel sync. Line and field oscillators. Output stages. Waveforms. Sandcastle pulse. D.c. power supplies.

Synchronizing pulse separators

The line and field *sync pulses* form an essential part of the complete TV signal, for they ensure that the image will be accurately reconstructed by the CRT. However, before these pulses can be used, they have to be extracted from the composite signal and then separated from each other. Since this operation immediately follows the video amplifier, it is important that these two stages should be well buffered or the input impedance will load the video amplifier and degrade its high-frequency response.

The basic principle involved in stripping the sync pulses from the composite video signal is depicted in Figure 4.1. The negative-going video signal is fed via C_1 into the base of the transistor that, without input signal, is held in the cut-off state. When driven by an input signal, the resulting base current causes C_1 to charge with the polarity shown. This provides automatic bias that holds the transistor conductive only during the sync-pulse period, thus producing a train of negative-going sync pulses as output. In order to provide stable synchronism, the time constant $C_1 R_1$ is normally made long compared with 64 µs. Noise pulses and interference that become superimposed upon the video can cause

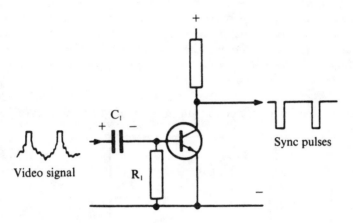

Figure 4.1 Sync-pulse separation

false synchronism by driving the separator stage into premature conduction. This can be avoided by using a *noise-gate* circuit of the type shown in Figure 4.2.

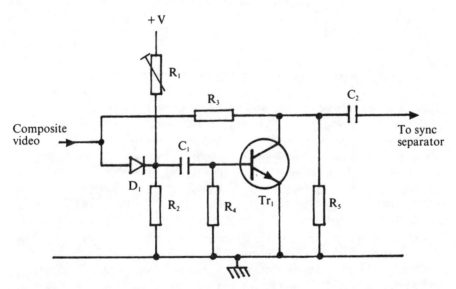

Figure 4.2 Sync-separator noise gate

The transistor Tr_1 is used as a switch in parallel with the feed to the sync-separator stage. Under normal conditions, Tr_1 is held cut-off by the bias provided from R_4. The diode D_1 further isolates this circuit, because it is normally reverse-biased. By suitably setting R_1 so that Tr_1 is held cut-off at levels about equal to the sync-pulse tips, any noise pulse of greater amplitude

73

causes the transistor quickly to saturate. This action short-circuits the sync feed into the separator stage and so removes the interfering noise. Because of the short time-constant of the circuit, the gate releases as soon as the noise pulse ceases. Both of these circuits can easily be incorporated into an integrated circuit.

The line and field sync pulses are separated from each other using differentiator and integrator circuits in the manner shown in Figure 4.3. These

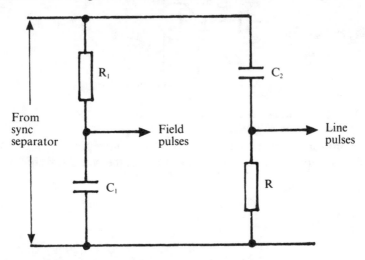

Figure 4.3   Separation of line and field sync pulses

circuits respond differently to pulses of different time duration. Typically the differentiator has a time constant of around 0.5 to 1 μs, which is short compared with the line-pulse duration of 4.7 μs. The output from this section is thus a series of very narrow line timing pulses. As shown in Figure 4.4 (and Figure 2.4 in Chapter 2), the field pulses consist of a series of five broad (27.3 μs) pulses which, because of the long time period, produce no effect on this circuit. Figure 4.4 shows how the inverted broad pulses are integrated in a circuit that has a time constant typically greater than about 40 μs. The output voltage *ratchets up* at each pulse to provide a large-amplitude output. By slicing this waveform at a suitable level, a broad field pulse of 2.5 lines duration is provided for field sync. If this sync pulse had been transmitted as a single element, line sync would be lost during field flyback, leading to erratic line sync at the beginning of each field. Because of the interlace requirement, successive fields differ by half a line, and so five line pulses are needed during the field-pulse period. So that odd and even fields are triggered at the correct time, it is necessary for the charge on the timing circuit to be the same at the start of the five broad pulses. Any error here would impair interlacing. The two sets of five

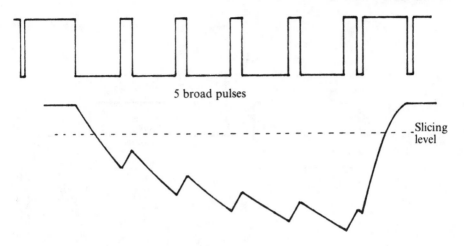

5 broad pulses

Slicing level

Figure 4.4   Recovery of field sync pulse

narrow *equalizing* pulses on either side of the broad pulses are therefore introduced to destroy any residual charge on the integrating capacitor.

Field timebase generators

The load for both timebase circuits is provided by the CRT scan coils. These have inductances and series resistances that are in the order of 3 mH and 3 Ω respectively. At the 50 Hz field rate (60 Hz in the USA), the inductive reactance is thus about 0.95 Ω (1.13 Ω), making the load impedance almost resistive and very similar to that found in many audio-amplifier systems.

Commonly, the field-timebase circuit consists of an oscillator, usually a multivibrator, which is triggered from the field sync-pulse sequence. The square-wave output is then used to charge a capacitor to produce the necessary ramp voltage and then discharged rapidly to complete the sawtooth. This waveform is then used to drive a Class A push-pull power amplifier stage, often in the totem-pole configuration, which in turn drives the field scan coils. The basic principle of this is shown in Figure 4.5. An overall negative feedback loop is included so that the wave shape can be linearized to compensate for the small inductive effect of the load.

A practical circuit

In Figure 4.6 Tr_1 and Tr_2 form a complementary-symmetry multivibrator circuit in which the two devices turn on and off together. This arrangement is often used because the circuit draws less power from the supply than the conventional

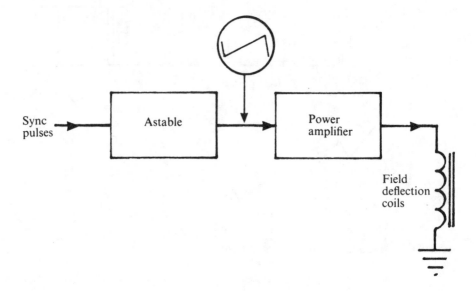

Figure 4.5 Basic field timebase generator

multivibrator. The speed of oscillations is controlled by C_1 and the resistance associated with the *hold* control. Initially, when both transistors are turned on, C_1 charges from the $+45$ V rail via R_1, Tr_1, D_1, R_3 and Tr_2. When C_1 is fully charged, Tr_1 turns off, and this lowers the base current through R_4 so that Tr_2 also turns off. C_1 now discharges slowly through R_5, R_6 and R_1, R_2. The voltage across C_1 falls to a point when D_1 becomes conductive causing both Tr_1 and Tr_2 to turn on again to repeat the charging cycle.

From the oscillograms it will be seen that the output is a series of short-duration negative-going pulses, coincident with the conduction periods of the two transistors. Positive-going sync pulses applied to the base of Tr_2 switch the transistors on to start the charging of C_2 and to synchronize the oscillator. Sawtooth waveshaping is produced by the circuitry between Tr_2 and Tr_3. C_2 charges slowly towards $+100$ V via R_8, C_3 and R_7 (D_2 and D_3 are cut off at this time). When Tr_2 turns on during flyback, D_2 becomes forward-biased so that D_3 discharges C_2 very quickly. The charging time for C_2 is controlled by the interval between sync pulses or by the charging time for C_1 if the circuit is unsynchronized.

As stated previously, the power-amplifier stage is a complementary push-pull arrangement commonly found in audio systems. It will be noted that a presettable negative-feedback loop is used to provide for overall scan-linearity adjustment.

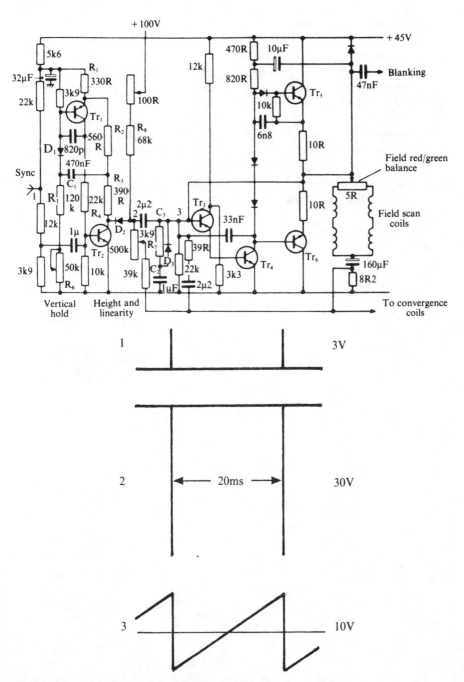

Figure 4.6   A field timebase circuit

77

Line timebase generator

The line timebase generator — comprising as it does the oscillator, driver, output stage and EHT generating circuits — is the most critical, the most complex and probably the most troublesome of any receiver, whether colour or monochrome. The reasons for this are as follows.

1 Because the line timebase is synchronized by a differentiated pulse, the arrival of any unwanted short pulse (e.g. of interference) can cause the timebase to fly back prematurely and impair synchronism.

2 At the line timebase frequency of 15.625 kHz, the typical 3 mH inductance of the scan coil produces an inductive reactance of about 295 Ω, with the result that the load acts as an *LR* integrator (time constant = *L/R* s). The sawtooth scan current can then be generated in the manner shown in Figure 4.7.

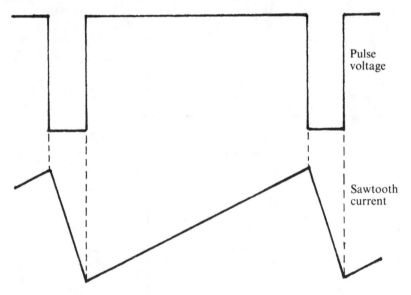

Pulse voltage

Sawtooth current

Figure 4.7   Derivation of scan current

3 The effect of the rapid flyback on the inductance in the line circuit is to generate a large e.m.f. to create insulation problems. (This large e.m.f., however, is put to good use in providing the high EHT required.)

4 The power consumption of the line output is high — in some cases as high as that of all the remaining circuits in the receiver put together.

5 Though free circulation of air round the line-output stage is required for cooling purposes, the stage must nevertheless be carefully screened to

prevent radiation of interference from the large-voltage pulses that occur within it (in the case of the colour TV receiver, screening against X-ray radiation due to the high voltage of 25 kV).

Flywheel synchronization

False triggering of the line oscillator is prevented by a circuit technique known as *flywheel synchronization*. In a flywheel sync circuit, every pulse from the sync separator, which may be either a true sync pulse or an unwanted interference one, is compared in a discriminator circuit with a pulse derived from the line output stage. The output of the discriminator circuit is a d.c. voltage whose amplitude depends on the phasing of the sync pulse relative to the line flyback. If the two pulses coincide — i.e., if the peak of the sync pulse occurs at the same instant as the mid-point of the (filtered) flyback pulse — there will be no change in the d.c. voltage output. If the sync pulse does *not* coincide with the mid-point of the flyback pulse, however, the d.c. output will alter. The change is used to correct the frequency of the line oscillator and synchronism.

Figure 4.8 shows a simplified flywheel sync circuit, in which the main discriminator portion is the bridge circuit formed by D_1-D_2-R_1-R_2. When the diodes D_1 and D_2 are both conducting, the voltage at the junction of R_1 and R_2 (which are resistors of equal value) is the same as that at the junction of D_1 and D_2. During the line sweep, the diodes are non-conducting, apart from a small bias current flowing into Tr_1 from the *line hold* potentiometer VR_1. When the sync pulse arrives from the sync separator, it is split into two pulses of equal and opposite polarity in a phase-splitter circuit, and these pulses are applied to the diodes. With the diodes then conducting, the voltage applied to the base of the transistor is the bias voltage from VR_1, plus any waveform derived voltage from the line output transformer through R_5.

If at this moment the sync pulse arrives in the middle of the flyback voltage, there will be no effect. If it arrives at any other time, the voltage may be either greater or smaller than the steady bias voltage, and the charge on C_4 will change accordingly.

The time constant C_4R_6 is made large enough to ensure that several successive pulses are needed to change the base voltage of Tr_1 appreciably. The collector current of Tr_1 then slowly changes from the value set by VR_1 until the sync pulses again coincide consistently with the mid-portion of the flyback pulse.

Tr_1 also functions as a reactance amplifier: i.e., a stage whose reactance varies with the applied d.c. just as the capacitance of a varicap diode does. In Figure 4.9, Tr_1 is connected as a common-base amplifier as far as signals at line frequency are concerned, because of the large value of C_1. It has a capacitor C_3 which feeds line signals from its collector to its emitter.

The emitter signal, however, is the voltage across R_3. When the reactance of

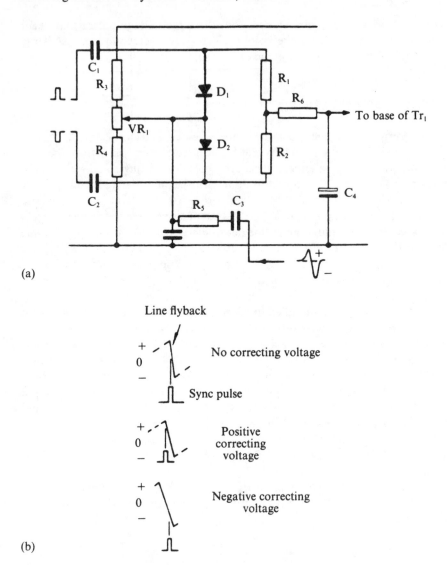

(a)

(b)

Figure 4.8   (a) The principle of flywheel synchronization; (b) the effect of correcting voltages

C_3 at line frequency is large compared to that of R_3, a phase difference of 90° will develop between the emitter voltage and the collector voltage. Since the emitter current is in phase with the collector current, there will also be a 90° phase difference between emitter voltage and collector voltage.

The transistor, in short, behaves like a reactor. The reactance (the ratio v/i)

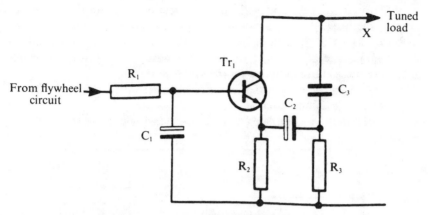

Figure 4.9   The reactance amplifier (simplified)

will depend on the current gain, which is in turn controlled by the magnitude of the base current flowing.

If the collector of Tr_1 in the reactance stage is connected into an oscillator circuit such as the Hartley type shown in Figure 4.10, the reactance of Tr_1 becomes part of a resonant circuit and so controls the frequency of the oscillator. The output voltage of the flywheel discriminator therefore controls

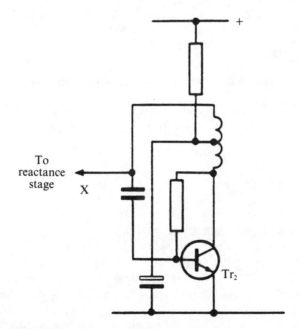

Figure 4.10   A Hartley sine-wave oscillator

the oscillator frequency, and will apply correction to it until the frequency becomes exactly equal to the sync-pulse frequency and in phase with the pulses. The flywheel part of the action is provided by C_1 which, by preventing rapid changes of voltage at the base of Tr_1 (in Figure 4.9) ensures that random interference pulses have no effect on the synchronization.

This action is in some ways analogous to the mechanical action of a flywheel: hence the circuit's common name.

Figure 4.11 shows a complete circuit of these last few stages. In it, Tr_1 is the

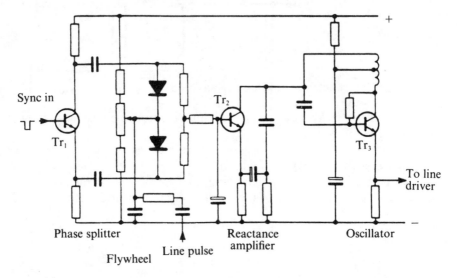

Figure 4.11　The complete flywheel line oscillator stage (simplified)

sync phase-splitter, having a transistor with load resistors of equal value in its collector and emitter circuits. Tr_2 forms the reactance stage and Tr_3 the Hartley oscillator circuit.

Driver and output stages

It has been shown that the line circuit is only required to generate a pulse voltage waveform at line frequency (Figure 4.7).

The line driver stage is a pulse amplifier which switches current into the line output transistor during the period of the sweep. A large current signal is needed, of the order of 1 A; so the usual method of drive is to insert a step-down transformer between the line-output transistor and the driver. The line driver operates at high voltages (typically boosted to 100 V) on the collector, and needs to be able to handle large currents also. A power transistor is therefore needed, fitted to an adequate heat-sink.

The line-output stage is always complex because of the many different tasks the stage needs to carry out. A specially designed transistor is called for, capable of withstanding collector voltages in excess of 1 kV and currents of several amperes (though never both of them simultaneously). The collector load of this transistor is a large inductor called the *line-output transformer* (LOPT). The scan coils are fed through a high-voltage type of isolating capacitor from the collector. The action can be followed by reference to Figure 4.12.

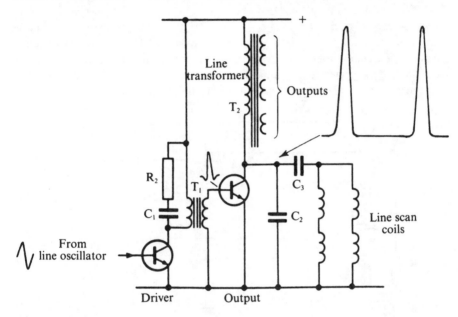

Figure 4.12 Action of the line-output stage

When the pulse from the driver stage is applied to the base of the line-output transistor, the latter is driven into full conduction. One end of the line-output transformer primary winding is suddenly earthed through the transistor, while the other end is connected to a supply at about 1 kV. Because of the inductance of the winding, full current does not flow immediately, but rises in a nearly-steady linear sawtooth form. The capacitor coupling causes current flow in the scan coils to change in a similar manner, but at a different d.c. level.

The action so far comprises the scan part of the waveform. During it, the voltage of the collector of the line-output transistor remains practically zero, though the current eventually reaches a value of several amperes as the end of the scan period approaches.

At this point, the driver stage switches over, reverse-biasing the base of the transistor. With the line-output transistor suddenly cut off, a large positive voltage is generated by the back e.m.f. caused by the collapsing field of the transformer primary. This charges the capacitor C_2, which then forms a resonant circuit with the inductance of the transformer. This resonant circuit performs a half-cycle of oscillation, so causing the flyback of the sweep.

The reason why the oscillation is damped out after half a cycle is because the collector voltage of the line-output transistor goes negative, so allowing current to flow between base and collector and thus shorting out the oscillations. With the additional windings, the stray capacitance of the line output transformer is likely to produce *ringing* (damped oscillations) that will continue well into the forward-scan period. It has been found that if the transformer leakage inductance is made to resonate at about the third or fifth harmonic of the line timebase frequency, there will be no energy stored in the inductance at the end of flyback. This prevents ringing and makes the system more energy-efficient.

Figure 4.13 illustrates the circuit of a line-output stage for a portable

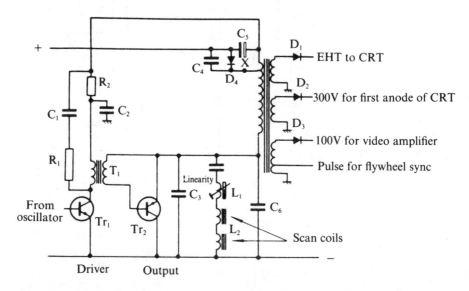

Figure 4.13 The line driver and output stage of a monochrome portable

monochrome receiver, in which Tr_1 is the driver and Tr_2 (a *p-n-p* type) is the line-output transistor. Several additional points can be noted from the diagram.

The diode D_4 and capacitor C_5 form an *energy recovery* or *boost diode* circuit. During the sweep portion of the cycle, current is drawn from the supply through D_4, but on flyback D_4 is reverse-biased, so that C_5 can charge to the full voltage at that point. This high voltage on C_5 is used at the start of the scan period to provide the supply voltage for the sweep, with D_4 conducting only when the voltage at *point X* falls below the level of the supply voltage (which in this case is 11 V).

The boosted voltage across C_5 is also used as the supply for the driver transistor (which in this design is 100 V).

The line-output transformer has several windings. One of these contains a large number of turns and can therefore produce a very large-amplitude voltage pulse of some 12 kV or more at flyback. This pulse is rectified and used to provide EHT for the CRT. Because of the high frequency of the line-output waveform, the output of the rectifier is easily smoothed, usually with the aid of the self-capacitance of a conductive coating on the outside of the CRT.

Another winding produces the supply for the first anode of the CRT while the output from a third winding produces the voltage for the video amplifier. A tapping from this winding is also used as the source of a blanking pulse, and to provide (after some wave-shaping) the output for the flywheel sync circuit.

Thyristor line output stages

Line-timebase circuits have been shown to be based on a switching action. Since thyristors are better than transistors in this respect they can be used instead with the following advantages.

1 Thyristors can pass high peak currents into low-impedance loads such as scan coils.
2 They have a very fast switching action.
3 They can safely withstand moderate *flash-over* voltages.
4 They produce a very low voltage drop in the on state.
5 Since they only require a very small gate pulse for turn-on, they are much easier to drive.

Figure 4.14 shows the basic principles involved in this variation. The switching action is handled in an alternate manner by two pairs of thyristors and diodes. These are usually incorporated into *integrated thyristor rectifier* devices. L_1 represents the scan-coil impedance that is coupled into the circuit by the line-output transformer and C_1 is the *S-correction* capacitor. With the deflection components D_1 or Th_1 on, the resonant frequency of the scan circuit is very low, controlled only by L_1 and C_1. When the deflection components are off, either Th_2 or D_2, the flyback components, are on and L_2C_2 is now added to the resonant circuit. While L_2 marginally increases the total inductance, C_2 consideably lowers the effective capacitance, with the result that the circuit has a much higher resonant frequency. This accounts for the sinusoidal appearance of the scan waveform.

Assuming that energy is already stored in the scanning circuit and starting at point A, D_1 is conducting, discharging this energy and causing the negative current to fall. At B, D_1 loses its forward bias and becomes cut off. Th_1 is now

85

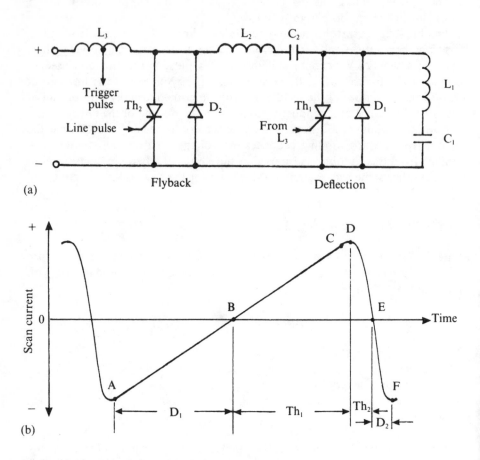

Figure 4.14   Principles of line thyristor timebase

triggered on by a pulse derived from the tapping on L_3 and via a phase-shifting network to provide the right-hand part of the line scan to D. At point C, about 3 µs before the peak, Th_2 is turned on by a line pulse. During the scanning period, C_2 has been charging via L_3 from the HT supply. When the flyback thyristor turns on, C_2 discharges to produce a current flow that cancels that flowing in Th_1. At D, the two currents completely cancel, so that the forward current through Th_1 falls below its holding value and so it turns off. As soon as Th_1 goes off, the current again begins to decrease but at a very much higher rate due to the higher resonant frequency. As the current starts to go through zero at point E, the drop in current through Th_2 causes it to turn off, but now D_2 becomes conductive to complete the scan cycle and the system is ready to restart at point A.

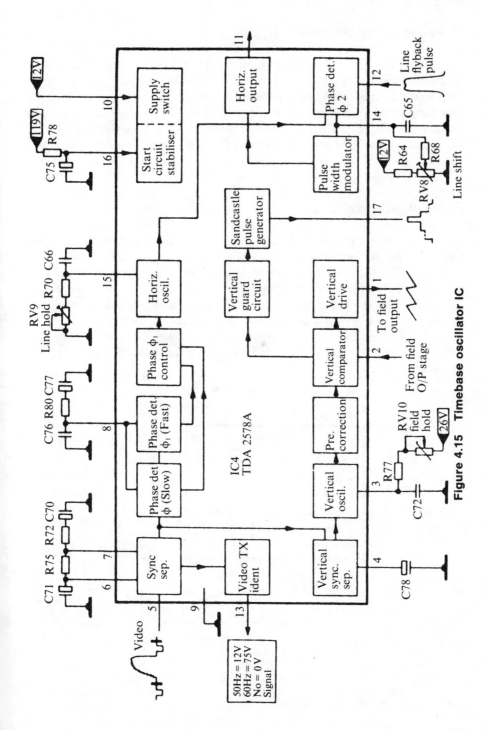

Figure 4.15 Timebase oscillator IC

87

Timebases using integrated circuits

Figure 4.15 shows the block diagram of the TDA 2578A, a highly integrated device used for timebase control. It has as its input the composite video signal, from which it extracts the field- and line-sync pulses to control the timing of the two oscillators. A flywheel-sync type of operation is provided for both oscillators, with the feedback from both timebase output stages being applied at pins 2 and 12. The hold controls operate by varying the d.c. potentials at pins 3 and 15.

Field output stage IC

Because this device (shown in Figure 4.16) operates from a relatively low 26 V supply, a flyback generator is incorporated. This produces a charge on capacitor C99, whose voltage is added to that of the supply to increase the scan capability. This chip provides enough power output to drive either 90° (TDA 3651) or 110° (TDA 3652) tubes. Apart from its own supply voltage stabilizer, the chip also contains a thermal-protection circuit. If the temperature rises above 175°C, the deflection current is reduced to lower the device dissipation. RV16 provides vertical picture shift by adding a d.c. level to one side of the field-scan coils.

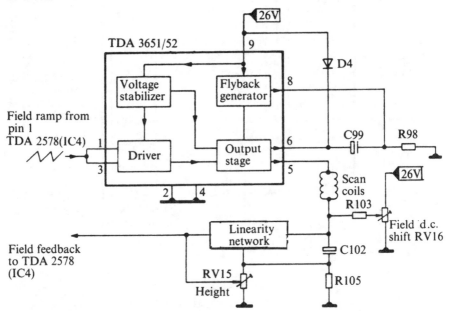

Figure 4.16 Field output stage

Horizontal generator and field timebase IC

This device (Figure 4.17) represents a further level in the process of circuit integration. Only the line power-output stage and a few additional components are needed to provide a complete scanning system. In addition, the field power-output stage has full built-in overload protection. The circuit is driven by composite video input and uses phase-locked loops (PLL) for oscillator control. In addition to scanning control, the chip also provides outputs for blanking (super sandcastle) and a video identification line that goes High to indicate that a valid composite video signal is being processed. This signal can be used to mute the audio system if necessary.

Line output stage and d.c. power supplies

The line output stage shown in Figure 4.18 follows the general principles already described, except that it uses the *diode-split* method for EHT generation. The usual large EHT transformer overwind is now split into three sections, each generating about 8.3 kV. These are coupled together with three high-voltage diodes that are now less heavily stressed than would be the case for a single very-high-voltage diode. In addition, the winding self-capacitance is used for the reservoir capacitor. This construction results in a smaller, more efficient and more reliable line output transformer.

Figure 4.18 also shows how it is possible efficiently to obtain many of the lower voltages required elsewhere in the receiver. The focus and A1 voltages are obtained via a potential divider from one 8.3 kV EHT winding section, whilst lower values can be obtained from separate windings on the LOPT. One particular advantage of this technique is that the high (15.625 kHz) operating frequency means that the ripple frequency is high, and can be filtered off using relatively small-value capacitors.

Servicing of line timebases

No servicing should ever be attempted on the line output stage until the manufacturer's service sheets have been consulted. The following general points should then be noted.

1 Voltage readings cannot easily and safely be taken at the line-output transistor collector because of the very high voltage that is present.
2 Oscilloscope measurements can be made on the base of each stage to establish whether the signal is reaching that stage or not.

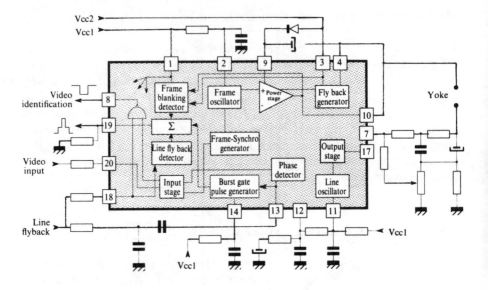

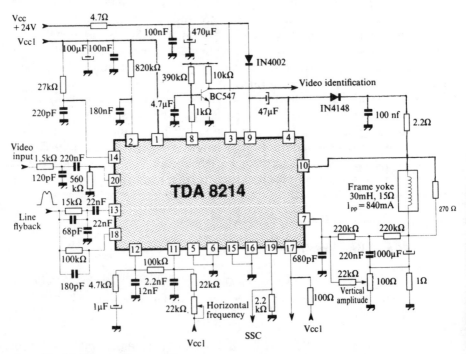

Note: Pins 5–6: Frame oscillator ground
Pins 15–16: Frame and line power ground

Figure 4.17 Horizontal generator and field timebase IC

90

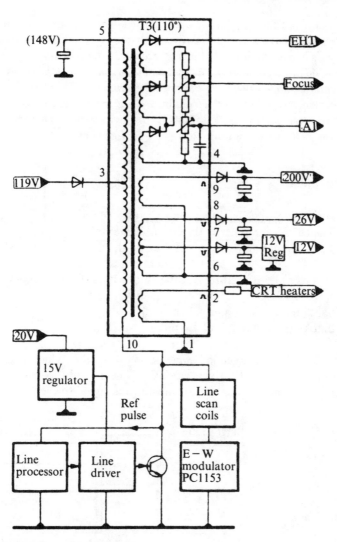

Figure 4.18   Line output stage

3   It is unfortunately rare for single faults to be found by themselves in a stage. A fault in one component usually has a 'knock-on' effect which damages other components. The most vulnerable of these are the line-output transistor, the line-output transformer (danger of internal shorts), the tuning and coupling capacitors (which must all be of the high-voltage-pulse working type), and the boost diode and capacitor. Failure of any of these components can cause failure in one or more of the others. This stage contains many safety-critical components and only replacement parts that

meet the appropriate standard should be used in any repairs. Failure to obey this rule could leave the repairer responsible for any subsequent damage to the owner's property, not only the TV receiver. *After a repair or replacement, therefore, all other components in the line-output stage must be checked before any attempt is made to operate the circuit again.*

4   The experience of fellow service engineers is a most valuable guide to fault symptoms. Articles on fault-finding in magazines like *Television* form an important part of the training of any TV service engineer.

Sandcastle pulse

In many application-specific integrated circuits (ASICs), pin-outs are at a premium. The sandcastle pulse is thus a good example of economy by design. Three blanking and gating pulses are needed in the colour TV receiver:

1   a narrow 4 µs pulse for gating-out the colour burst from the back porch to synchronize the colour sub-carrier frequency;
2   a line-blanking pulse of about 12 µs; and
3   a field-blanking pulse of about 1.4 ms.

Figure 4.19 shows the so-called *super sandcastle pulse*, which is commonly derived within a line-and-field generator IC. The tri-pulse combination thus only requires one pin and one line to transfer this signal to the luminance and

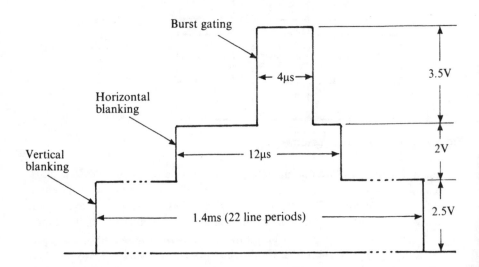

Figure 4.19   The sandcastle pulse showing nominal values

chrominance processing chip where it will be separated out using three level detector stages.

Horizontal and vertical controls

The controls associated with both timebases are usually of the pre-set type with non-user access. To obtain best results, service adjustments that affect picture linearity should be made using a suitable test pattern as receiver input.

The field or vertical-hold control normally adjusts the HT supply to the oscillator stage, which varies the charging period of the timing capacitor.

The line or horizontal hold varies the bias on the flywheel sync stage and in addition may include an inductor-core adjustment. Both of these circuits should be adjusted according to manufacturer's information, and this usually dictates that the sync feed should be disabled. However, a useful *rule-of-thumb method* is to rock each control separately to the left and right, noting the points at which synchronism slips, and then position each control midway between these points.

Height adjustment is normally affected by a pre-set resistor in the field-linearity feedback circuit. The line width is achieved either by adjustment of the level of HT supply to the line circuit, or by the use of a *saturable inductor* in series with the line-scan coils. (A saturable inductor has a value of inductance that varies with the current flowing through it.)

Line- and field-linearity controls may be included. The first may be a second saturable inductor in series with the line-scan coils; field-linearity adjustment is almost always made via the feedback network.

Line- and field-shift controls, that inject a small level of d.c. into the scan coils, are used to centre the picture within the CRT mask.

Geometric distortion

The displayed raster, particularly on large, wide-angle deflection tubes, becomes distorted in the manner shown in Figure 4.20. This is described as *pincushion distortion*, which occurs basically for two reasons:

1 equal changes of scan current produce equal changes of beam deflection angle;
2 the beam deflection centre does not coincide with the tube face radius.

For similar reasons, a linear sawtooth line waveform also produces distortion. This can be corrected by adding a suitable capacitor in series with the scan coils to give the waveform the S shape shown in Figure 4.21. The S correction capacitor is C137 in Figure 4.22.

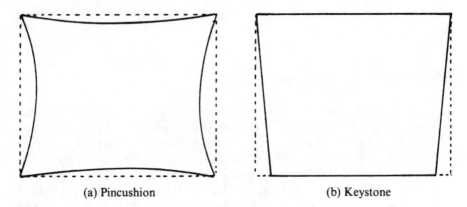

(a) Pincushion   (b) Keystone

Figure 4.20   Geometric raster distortion

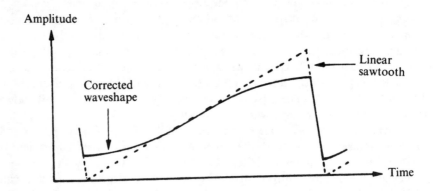

Figure 4.21   S-corrected line sawtooth: scan correct

Since the picture height is less than its width, the north–south (NS) distortion is less than that in the east–west (EW) direction. In a monochrome tube EW correction is commonly applied using small permanent magnets mounted close to the scan coils. These assist the line-scan current, particularly across the middle of the screen. Often no correction is provided for the small NS distortion in the monochrome case. Small permanent magnets may be used with a colour TV tube, but alternative methods are more common. Some correction can be made using scan coils specifically designed for a particular tube type. Other methods involve modulating the line scan current with a field-scan signal and vice versa. Figure 4.22 shows a popular method using the *diode modulator*, together with its necessary preset controls. Over-correction of the pincushion leads to *keystone* distortion as shown in Figure 4.20(b). This may be corrected by injecting an anti-phase sawtooth signal into the scan coils via a keystone control.

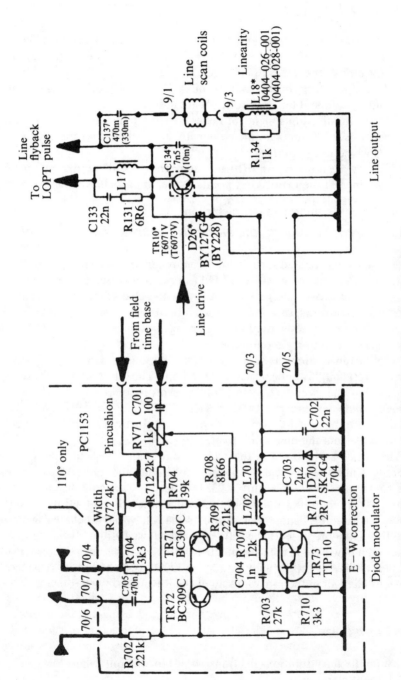

Figure 4.22 Line-output stage and EW correction circuit

Exercise 4.1

(a) Using an oscilloscope and test-pattern generator, trace and sketch the timebase waveforms for field- and line-scan systems (excluding the collector of the line-output stage).

(b) Note the effects on the waveforms and the CRT display as the height/width, hold and linearity controls are adjusted.

(c) Carry out the adjustments of raster-correction magnets and the geometry controls in a monochrome/colour TV receiver, noting the effects on the displayed test patterns.

(d) Repeat sections (a) and (b) but note and record the d.c. voltage levels obtained at the various test points. Note, in particular, how these vary as the controls are mis-adjusted.

Exercise 4.2 Testing of line output transformers and scan coils

Even one short-circuited turn on an inductor can seriously reduce its efficiency. When such a condition exists on a LOPT or scan-coil assembly, the effect can be disastrous. D.c. ohmic tests will not reveal the change of resistance for even several shorted turns, thus making such a fault very difficult to identify. Often, after making many tests, the LOPT is changed in desperation. Such diagnostic techniques are not very economical.

When a tuned circuit is shocked into oscillations, a damped sine wave that decays exponentially appears across the circuit. The time taken for the decay is a function of the circuit Q factor, and this property is used in the so-called *dynamic inductor testers*. These devices allow many LOPT and scan-coil assemblies to be tested *in circuit* with power off, by isolating only the EHT and focus leads from the timebase circuit.

The principles involved in this technique provide a very instructive training exercise. Calibrate an oscilloscope on both voltage and time scales. Connect up the circuit as shown in Figure 4.23, set the timebase to 2 ms/cm and the X sensitivity to about 0.3 V/cm. Set sync to '+ internal' and adjust for a steady trace. The circuit is being shocked into oscillation by the timebase trigger pulses and the display represents the train of damped oscillations. If this test is being made on a new LOPT, then a simulated short-circuit turn can easily be made by looping a piece of copper wire around the LOPT core. The result produces very obvious changes in the rate of decay of the waveform.

Test questions

1 State the time durations (as transmitted) for the following:
   1 line-sync pulse;

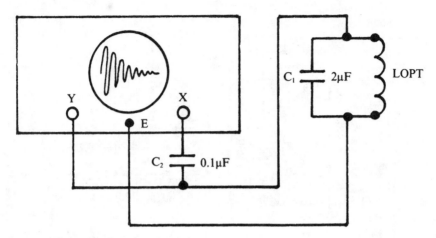

Figure 4.23   LOPT testing (Exercise 4.2)

1 broad-field pulse;
1 equalizing pulse;
the front-porch period;
the back-porch period.

2   Refer to Figure 4.24.
   (a)   What is the significance of the triangular symbol shown against R97?
   (b)   Identify the field height and linearity controls.
   (c)   Why are the field-scan coils parallel-connected?
   (d)   What signals are carried over the lines A and B?
   (e)   What is the expected shape of the waveform at test point 12?

3   Refer to Figure 4.25.
   (a)   Identify the S-correction capacitor.
   (b)   Identify the line-width and linearity controls.
   (c)   What function is performed by D26?
   (d)   What signals are carried over lines A and B?
   (e)   What noticeable effect would occur if D24 became short-circuited?

4   Refer to Figure 4.26.
   (a)   State the function performed by the circuit associated with TS800.
   (b)   What purpose is served by the output feed B?
   (c)   Identify the line-scan coils.
   (d)   What is the function of C622?
   (e)   What function is performed by Block C?
   (f)   Identify the height control.

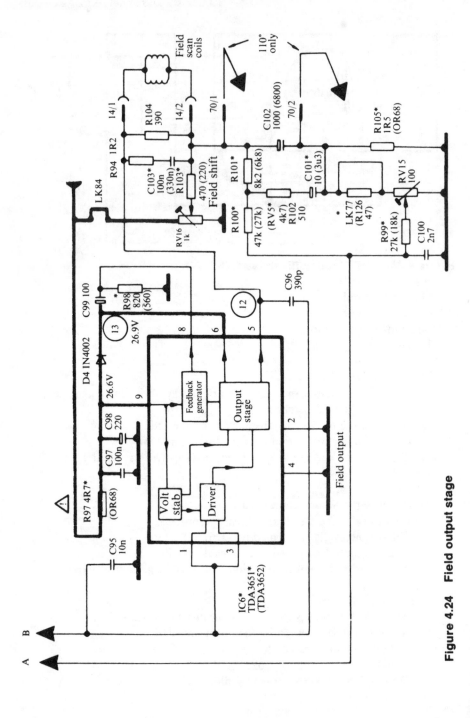

Figure 4.24 Field output stage

98

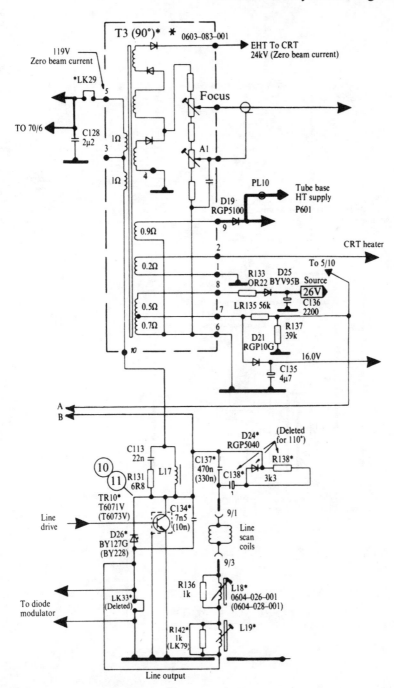

Figure 4.25   Line output stage

5 Cathode-ray tubes, associated circuits and liquid crystal displays

Summary

CRT construction and principles of operation. Deflection techniques. CRT applications. Operating conditions. Video-drive amplifiers. Grid or cathode drive. Convergence. Degaussing. Servicing and setting up. LCD display technology and scanning control.

The cathode-ray tube (CRT)

The cathode-ray tube operates by harnessing the movement of electrons in a vacuum, on the following principles:

1 Electrons are released into a vacuum when a cathode material, such as barium oxide, is heated to 1000°K or more.
2 Electrons, being negatively charged, are attracted to positively charged metal plates and are repelled from negatively charged plates.
3 Moving electrons constitute an electric current, and so can themselves form part of a circuit.
4 The direction in which a stream of electrons moves can be changed by applying to it one or more electromagnetic or electrostatic fields.
5 Moving electrons carry energy, and so will cause substances called phosphors (which have no connection with phosphorus) to glow brightly when they are struck by an electron beam.

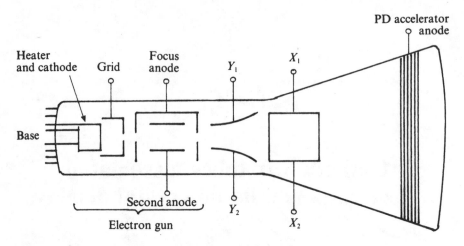

Figure 5.1   CRT for instrumentation applications

Cross-sectional diagrams of two typical CRTs are shown in Figure 5.1 (an instrument tube) and Figure 5.2 (wide-angle tube used in black-and-white television). The function of the various parts is as follows, and is illustrated in Figure 5.3.

The cathode is a nickel cup, coated at its closed end with a mixture of oxides of barium, strontium and calcium, which emit electrons when at red heat (Figure 5.3(a)). Heating is carried out electrically, using a molybdenum or tungsten wire which has to be coated with aluminium oxide to insulate the heater from the cathode. Failure of this insulation is a fairly common CRT fault.

Surrounding the cathode is another, larger, metal cup called the *grid*, or *control grid*. This has a small (0.025 mm or less) hole at its centre. The heater, cathode and grid constitute the source of the electron beam and require four connections: two for the heater, and one each for the cathode and grid.

The voltage between the cathode and the grid controls the flow of electrons through the hole in the grid. In the cut-off condition, typically at a grid voltage of about 85 V negative with respect to the cathode, no electrons pass through the hole in the grid. As the grid voltage is increased positively to about −5 V relative to the cathode, the electron beam current increases progressively to maximum.

Operation of a CRT with the grid *positive* to the cathode is normally undesirable, but it can help an elderly tube to sustain electron emission, and is a technique sometimes used in conjunction with higher-than-normal heater currents, in CRT reactivation.

The electron stream emerging from the hole in the grid is a diverging stream which then needs to be converged, or *focused*, so that a spot of light is produced at the exact point where it hits the screen. Focus is achieved by placing, at appropriate places along the length of the gun, metal cylinders (anodes) carry-

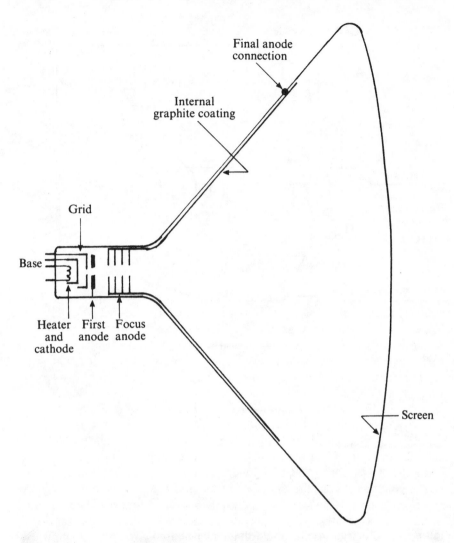

Figure 5.2   CRT for monochrome reception

ing different voltages. This system is called *electrostatic focusing*. It is more commonly used nowadays than an alternative technique called *electromagnetic focusing*, in which a permanent magnet or a focus coil through which d.c. is flowing is positioned along the neck of the tube. Since the focus must be adjustable, some provision has to be made for altering either the voltage of the focus cylinder, the position of the permanent magnet, or the current through the focus coil.

Once focused by any of the methods mentioned, the beam has then to be

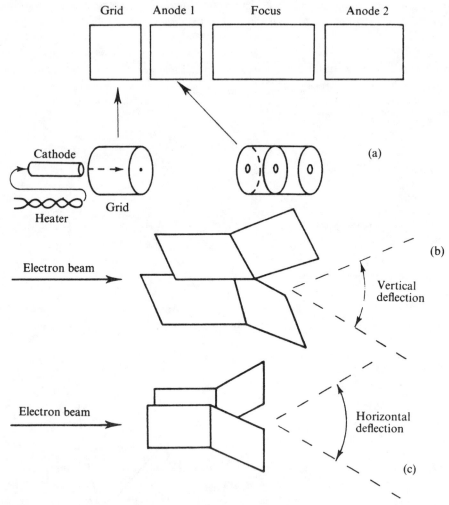

Figure 5.3   Electron gun for electrostatic deflection

deflected so that it can be directed at any part of the screen. Electrostatic CRTs used in oscilloscopes and miniature TV receivers use metal deflector plates to achieve this. The pair of plates situated closer to the cathode are called the *Y-plates*, and are used for *vertical* deflection of the beam (Figure 5.3(b)). The other pair, mounted at 90° to the plane of the *Y*-plates, are called the *X-plates*, and cause *horizontal* deflection of the beam (Figure 5.3(c)). Deflection sensitivity, measured as the number of millimetres of spot deflection per volt difference between the plates, is greater for the *Y*-plates than for the *X*-plates.

In TV and radar, magnetically-deflected CRTs are used, and there are no internal deflecting plates. The magnetic deflection coils form an external component located around the neck of the tube, where they are supplied with current signals.

Finally, after focusing and deflection, the electron beam strikes the phosphor material deposited on the screen. This material, which is an insulator, is generally coated with a thin film of aluminium which acts in two ways:

1 it provides a metal contact so that a high (positive) accelerating voltage can be applied to it; and
2 it reflects light from the phosphor which would otherwise be lost into the tube.

Electrons can penetrate this aluminium layer quite easily — a fact which makes the gains to be had from using such a layer easily exceed the losses.

The voltage which is applied between the cathode and the screen (the *final anode connection*) is called the *EHT* (which stands for extra high tension). It has the effect of accelerating the electrons towards the screen. The greater this acceleration, the brighter will be the spot when the full beam current strikes a phosphor dot on the screen.

The value of the accelerating voltage also has an effect on the deflection sensitivity. At large values of EHT, much more deflection effort, be it voltage between plates or current through coils, is needed to achieve the same number of millimetres of deflection than is needed at lower EHT values. Electrostatically deflected tubes suffer rather more from this problem than do magnetically deflected CRTs.

The choice of deflection method

Instrument CRTs invariably employ electrostatic deflection. This method of deflection is well suited to signals which can range in frequency from d.c. to a few hundred MHz, because the deflection plates behave in a circuit as a high impedance with a small-value capacitance. Since no steady current is drawn by the plates, they can be supplied with signals from a voltage amplifier.

The use of plates for deflection limits the amount of EHT which can be used, because large EHT voltages reduce deflection sensitivity to unacceptably low values. Bright traces, which are needed when very fast deflection is required, are obtained by further accelerating the electron beam *after deflection*. The technique is called *PDA*, or *post-deflection acceleration*. A CRT using this technique operates with a second/third anode voltage of around 1000 V and a PDA anode voltage at least twice this value.

Television and radar CRTs need screens of much larger size than do instrument tubes so much higher beam currents are needed to produce an acceptable level of brightness over the whole screen. Acceleration voltages of 8 kV to 25 kV are common, depending on the size of the tube and whether the set is monochrome or colour, the highest accelerating voltages being required by large colour tubes.

For such high beam currents and acceleration voltages, the technique of magnetic deflection is appropriate because the deflection frequency itself is either fixed or only slightly variable, so that the coils can be designed for optimum operation at that frequency.

Typical operating voltages

Table 5.1 below shows a typical range of voltages encountered on the CRT

Table 5.1

	Cathode	Grid	First anode	Focus anode	Final anode
Monochrome	+80V	+20V	+300V	up to +350V	10 – 15kV
Colour	+150V	+40V	+1500V	up to +4kV	25kV
Oscilloscope	−100V	0V	+100V	0 to −200V	2 – 4kV

electrodes used in different applications (in practice values may differ significantly from those quoted).

In the oscilloscope it is convenient to use a combination of negative and positive supplies to obtain the high potential required between the cathode and the final anode. This allows the grid electrode to operate near to earth potential.

Electrodes and voltages

With the exception of miniature receivers, the CRTs used in all British TV receivers are magnetically deflected. The electrode structure of the gun is therefore comparatively simple, and only brightness and focus voltages need to be adjustable.

The connections to a typical monochrome tube are as follows:

1 *Heaters*, generally 6.3 V or 12 V.
2 *The cathode*, to which the video signal is taken from the video amplifier, is usually at the d.c. level of the collector of this amplifier (typically about 80 V).
3 *The grid*, decoupled to signal frequencies but at a d.c. level which can be controlled to provide control of brightness. The grid is sometimes also used to inject blanking (i.e. black-out) pulses into the signal, in which case it is not decoupled. Some receivers control brightness level by means of the cathode voltage, in which case the d.c. level of the grid is fixed (typically about 20 V).
4 *The first anode*, which is always kept at a fixed d.c. voltage (typically about 250 V).

5 *The focus electrode*, run at a d.c. voltage which can be varied to give control of the size of the spot on the screen (typically 200 to 400 V).
6 *The final anode*, which works at the full EHT voltage. This voltage is typically 10 to 15 kV, depending on the size of the screen, *and is dangerously high to anyone probing the interior of a TV set while it is switched on.*

Colour tubes have three separate electron guns with connections similar to those described for the monochrome tubes. However, the electrode voltages are somewhat higher. Typical values might be around 150 V for cathode, 40 V for grid, 1500 V for first anode, 4 to 6 kV for focus anode and 25 kV for final anode.

The *scanning yoke* of the coils which fit round the neck of the colour tube is much more complex than is the case on the monochrome tube.

Never attempt to measure EHT with an ordinary meter. While most other voltages within a TV receiver can be measured with a general purpose multi-range meter, the EHT can only be measured with a special high-impedance electrostatic meter.

Tube replacement

The procedure for replacing a faulty or damaged tube varies somewhat from one receiver to another, and the manufacturer's handbook should always be consulted. The glassware is most vulnerable at the tube neck and where the glass changes shape most rapidly. The *rim band* not only provides a point of common earth and a means of mounting; it also provides a restraining force against the expanding force caused by the air pressure on the face plate. On a 60 cm diagonal tube, this represents a force equal to a weight exceeding 2000 kg. If the tube is damaged it may therefore implode and cause glass fragments to fly at dangerously high speeds.

The following general notes will nevertheless be useful.

1 Always wear safety goggles and gloves to avoid injury from flying glass, should the tube implode.
2 Place the receiver on a clean bench, with sufficient space available to lay the CRT alongside it once it has been taken out. Cover the bench surface where the receiver and CRT are to be laid with a blanket.
3 Disconnect the receiver from the supply, and discharge all capacitors. The outer coating of carbon on the tube, which is often used as one of the plates of a smoothing capacitor for the EHT, must also be earthed at all times.
4 Remove carefully the tube base, the EHT connector, and the leads to the scanning assembly. Take off the scanning yoke, if applicable, after noting carefully its exact position and orientation against the time of re-assembly.
5 Turn the receiver over on to its face and carefully remove the tube clamps. Lift the tube out, holding the ends of the screen and protecting the neck of the tube from any impact. Place the tube on the blanket, face down. *Never* lift a tube by its neck.

6 Observe all safety precautions until the tube is packed into a crate and covered.

CRT circuitry

Figure 5.4 shows some of the circuitry on and around the CRT of a portable

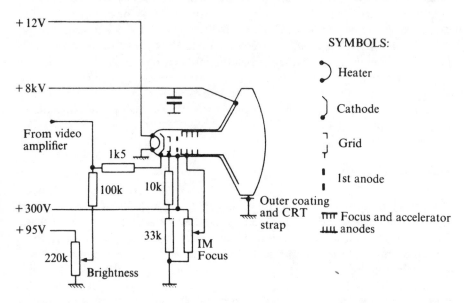

Figure 5.4 Some of the circuitry serving the CRT of a portable monochrome television receiver

monochrome receiver. In this particular circuit, the 12 V supply for the tube heater is obtained from the d.c. line which is also used for most of the transistors in the circuit. For mains-operated receivers, this supply is commonly derived from a winding on the line output transformer (see Chapter 4).

The cathode of the CRT shown is supplied from the collector of the video output transistor, with a 1k5 resistor between them to limit the amount of current that can flow. Spark gaps are provided at strategic points in the circuit to protect external transistors from the transient effects due to *flash-over* within the CRT. Such conditions, which may rupture the semiconductor junctions, can arise from sudden changes in the level of the EHT voltage or from sharp changes in the brightness level due to picture information.

In the design illustrated, the *brightness* control is governed by the d.c. level of the cathode, and the grid circuit is used for blanking. Blanking pulses are negative pulses applied to the grid of the tube to ensure that the electron beam is cut off when no picture signal is present (as during line and field flyback). Though the structure of the signal itself should ensure this cut-off, additional blanking provision is generally necessary.

The first anode of the tube is supplied with about 300 V from a half-wave rectifier circuit, using power obtained from the line timebase (see Chapter 4). This same supply is used also for the focus electrode. The current taken by the focus electrode is small, so that high resistances can be used in this part of the circuit.

The final anode is fed with EHT from another half-wave rectifier, also supplied from the line timebase stage. The EHT voltage used in this particular design is much lower than usual — only about 8 kV. The EHT supply is

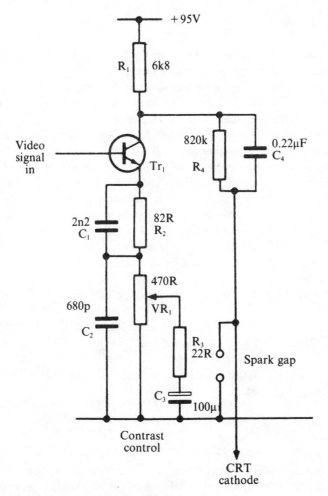

Figure 5.5 The video amplifier stage

therefore simple. Colour receivers generally need multiplier circuits to step up the EHT voltage, with some method of stabilizing the voltage also incorporated (see Chapter 4).

In the video-amplifier stage of the portable monochrome receiver illustrated in Figure 5.5, contrast control is achieved by varying the gain of the video output transistor Tr1. In the circuit illustrated, this is done by using a variable-emitter bypass for signals only. D.c. passing through Tr1 flows also through VR_1, the *contrast* potentiometer; but some of the signal current can be bypassed through R_3 and C_3. The gain is highest (so giving the greatest picture contrast) when the slider of VR_1 is set at the end of VR_1 which is connected to C_1 and C_2.

Grid or cathode modulation

Under normal conditions the grid is biased negative with respect to the cathode (k). The beam current may be increased either by driving the grid positively or the cathode negatively. Figure 5.6 shows the grid-voltage beam-current characteristic for an arbitary CRT. Assume that the voltage between anode and cathode (V_{a-k}) is initially set to 300 V (cathode at 0 V) when the grid is biased to -55 V. A positively increasing signal of 50 V peak will progressively increase the beam current to a maximum at a grid value of $-5V$. If an inverted version of the same signal was applied to the cathode, the beam current would again increase to a maximum when the cathode reached -50 V. As the cathode voltage swings negatively, the anode to cathode voltage is increasing. When the cathode reaches -50 V (maximum beam current), the anode-to-cathode voltage has progressively increased to 350 V. The CRT thus operates dynamically along the steeper dotted curve of greater sensitivity. This means that lower amplification is needed, and this can be achieved with a lower value of collector voltage for the video-driver stage.

Further advantages accrue from the use of cathode modulation as follows.

1 The cathode presents a low input impedance to the electron beam as opposed to the very high impedance found at the grid. The input signal line is therefore very much less susceptible to interference.
2 Since d.c. signal coupling can be applied to the cathode, the low-frequency response is improved.

CRT adjustments

The EHT voltage to a monochrome tube is generally not adjustable. Where adjustment is possible, as in the case of earlier colour TV receivers, *the manufacturer's instructions must be followed exactly*. In recent models, this operation is

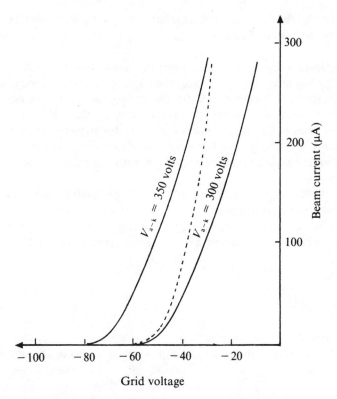

Figure 5.6 Grid-voltage/beam-current characteristic

carried out indirectly by the setting of the HT line voltage to the line-output stage.

When adjustment of contrast, focus or brilliance is required, the receiver should first be allowed to warm up, and then made to show a test pattern to check picture focus, and a grey scale to check the contrast range. The procedure is then as follows.

1 Use the *colour* control (on a colour receiver) to turn off the colour, leaving only a monochrome picture on the screen.
2 Set the *contrast* control to minimum contrast.
3 Adjust the *brilliance* control so that the black areas of the picture are *just* at an acceptable black, and set the *contrast* control so that there is a good contrast ratio between blacks and whites, with all the stripes of the grey scale clearly distinguishable. Some further slight adjustment of the *brilliance* control may then be needed.
4 Adjust the *focus* control so that close-spaced vertical lines can be seen. The degree of resolution which is obtainable here is governed by the bandwidths

111

of the IF and video stages, so that adjustment of the *focus* control itself may only have a limited effect.

A correctly adjusted picture should display black areas which are truly black with no scanning lines visible in them, clear highlights, and a good range of grey tones between these extremes. A dull or dirty-grey picture indicates either insufficient contrast or a low-emission cathode in the CRT, which should probably be replaced soon; a so-called 'soot-and-whitewash' picture indicates excessive contrast. Too low a setting of the *brilliance* control causes detail in the grey areas to be lost, while too high a setting results in a picture lacking areas of true black.

Viewers generally tolerate pictures of abominable quality, and some will even alter a correctly-set picture so as to make it conform to their own subjective standards. But it is the job of the service engineer to ensure that no receiver leaves his workshop before the picture it will give is correctly set up and adjusted.

Colour TV tubes

Figure 5.7 shows the construction of a modern colour TV tube. The three electron guns, one for each primary colour, are now placed *in line* and parallel to

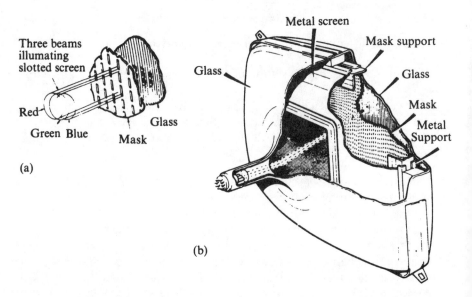

Figure 5.7   Colour TV tube: (a) cut-away section of mask; (b) cut-away view of picture tube

each other, in the horizontal plane. (In earlier — delta gun — tubes, the guns had been arranged in a triad.) The three beams are deflected together, under the influence of the scan coils, mounted as usual on the tube neck, to impinge upon the phosphor stripes on the inside of the tube faceplate. A slotted *shadow mask* matched to the phosphor positions during manufacture is used to ensure as far as possible that each beam only activates its own colour phosphor. The actual colour radiated by a particular small area of the tube face then depends upon the intensities of the red, green and blue beams illuminating that area.

When the beams are deflected away from the central area of the screen, they will mis-converge because of radius differences between the beam deflection and the tube face, and other small geometry discrepancies. This results in a display with coloured fringes. The three beams can be *converged* by applying correcting magnetic fields. Since the guns are in the same horizontal plane, the mis-convergence in the vertical or field direction will be very small. In early versions of the *precision-in-line* (PIL) tube, some correction for mis-convergence was provided by multi-pole magnets mounted on the back of the scan coils. Modern versions of the in-line tubes do not need any convergence control. This is because the scan coils have been constructed so that their characteristics match those of the CRT. The scan-coil assembly can then be fixed with a thermosetting adhesive during the last stages of manufacture. Other tubes have to be converged after certain service operations. If the convergence operation is to be carried out successfully, the method stated in the manufacturer's data sheet must be strictly followed.

The Trinitron tubes produced by Sony have three separate in-line cathodes to produce the electron beams. The outer two cathodes are slightly inclined inwards so that the three beams cross each other before passing through a beam-separating lens that causes re-convergence at a single point on the tube face. Between these two stages, the three beams are processed through common accelerating anodes. The shadow mask in these tubes is replaced by a vertical aperture grille structure that achieves the same effect.

Under normal operating conditions, the shadow mask and its associated metal work can develop a low level of permanent magnetism. This is usually removed by the action of the degaussing coils (mounted on the tube bulb) every time the receiver is switched on. From time to time, this magnetism may build up to produce coloured patches that are most obvious on a blank white raster. The metal work must then be *degaussed* (demagnetized) and the *purity* controls reset, to ensure that each beam only activates its own colour phosphor over the whole screen. Degaussing is carried out by exposing the whole of the tube face to a slowly decaying a.c. magnetic field using a degaussing coil. The purity adjustment is best carried out on a single colour, usually with only the red gun on, as this makes unwanted coloration more visible. Proceed by sliding the scan coils backwards and adjusting the purity rings until a red area is displayed in the middle of the screen. Slide the scan coils forwards until the screen displays a

113

completely red raster. Lock up the scan coils. Like convergence, this operation is no longer needed with certain in-line tubes. Purity problems are otherwise introduced by mis-convergence, deflection defocusing or astigmatism (image distortion).

Beam current limiting

Under normal operating conditions, the beam currents in the colour tube must be limited to some specified value. Excessive current can cause the shadow mask to overheat and distort, leading to serious and permanent purity errors. In a slightly less serious way, it can also cause an overload of the line output transformer (LOPT), reduced EHT, poor picture quality and errors of convergence, purity and focus. Limiter circuits sense the level of the beam current and generate a control voltage. This is then used either to reduce the brightness or contrast, or to lower the level of the power supply.

CRT faults

The following are some typical faults:

1 *low maximum brightness*, the causes of which could be either low EHT, faulty voltage supply to the grid or low cathode emission (in an old tube);
2 *uncontrollable brightness*, caused by shorts between either heater and cathode or grid and cathode;
3 *deflection irregularities*, usually caused by amplifier failures, but if an instrument tube is dropped or sharply knocked, a deflection plate can be loosened or even detached;
4 *no trace*, probably caused by amplifier or power supply failure, or by an o/c tube heater;
5 *other tube faults*, including misaligned grid, o/c connections, and leakage of air into the tube. A leaking ('down-to-air') tube can be detected if the *gettering* (the name used to describe the normally silvery deposit at the neck of the tube) has turned white.

Liquid crystal displays (LCDs)

Liquid crystals (LCs) are so called because their molecules are free to move about as in a liquid, but are arranged in an orderly manner like crystals. These properties are temperature-dependent and at about $-10°C$ and $80°C$ they become solids and normal liquids respectively. The molecular structure is also influenced by electric fields which must be generated by alternating voltages, typically between 30 Hz and 1 kHz (d.c. voltages produce electrochemical actions which degrade the life time of the devices). The *twisted nematic* (TN) liquid crystals used for display panels consist of rod-shaped molecules organized in a tightly twisted form. When influenced by an electric field, the structure untwists.

Display panels are made by sandwiching the LC between two plates of glass, spaced by about 6μm, that carry indium tin oxide patterns to provide the conduction paths for the electric fields. The structure is further sandwiched between two transparent light polarizers which are arranged for cross-polarization. Light energy incident on one side of the panel will decreasingly pass through the structure as the molecules untwist and align with the electric field. This effect is produced by an increasing electric field strength, as shown in Figure 5.8. If one of the polarizer plates is rotated through 90°, the reverse action takes place.

For the small display panels used for TV receivers, the conducting paths are deposited as a matrix of *row* and *column* lines on the inside of the structure.

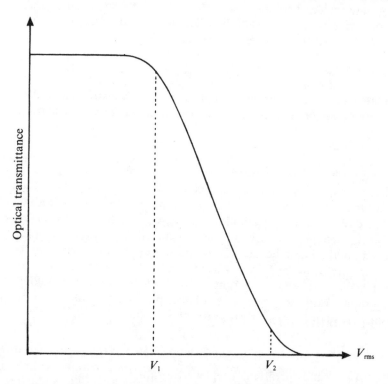

Figure 5.8   Transmittance of TN-LCD using crossed polarizers

When a *fluorescent* back-light source is added, the selection of a particular row and column line causes one specific picture element or *pixel* to be illuminated. For TV applications, it is necessary to cater for a grey scale, and for this it is necessary to vary the amplitude of the time-averaged r.m.s. voltage applied to the address lines. This can be achieved by varying the applied voltage between the values of V_1 and V_2 as shown in Figure 5.8 using a multiplexing technique that generates both the selection and contrast voltages.

Two important improvements on this basic structure are employed in the more recent panels. Each pixel is selected and illuminated via a *thin-film transistor* deposited at each row and column intersection to give better control over the grey scale. By using a liquid crystal with the so called *super-twisted* (super TN) molecular structure, a wider viewing angle is provided.

Colour TV panels are provided by depositing suitable red, green and blue filters over each group of three pixels. Thus when the LC is illuminated, the appropriate colour shines through each pixel.

Raster scanning and pixel control is achieved by using a controller chip built into the panel. This converts the line and field sawtooth scan voltages into appropriate row- and column-selection pulses. At the same time, a pulse whose amplitude represents the level of the video signal is added to provide the correct degree of contrast for the selected pixel.

Exercise 5.1

Examine an assortment of electron guns from both instrument and television CRTs, and note the position and structure of the various electrodes.

Exercise 5.2

Use a monochrome TV receiver with no input signal.

With all supplies to the tube switched off and all leads shorted one after another to earth in order to discharge any capacitors, attach the leads of a high-resistance voltmeter to the cathode (+) and grid (−). Replace the covers so that no points where high voltages will be present can be touched when the tube is switched on. (It should be made *impossible* for you or anyone else to touch any part of the voltmeter leads save where they are properly covered by insulation.)

Switch on, and note the action of the brilliance (*brightness*) control. If a photographic exposure meter is available, plot a graph of screen brightness (representative of beam current) against bias voltage.

Exercise 5.3

(a) Carry out the operation of replacing a cathode ray tube.
(b) (i) On a suitable colour TV receiver, carry out the degaussing operation and purity adjustment.

(ii) Using a pattern generator, carry out the convergence operation, following the manufacturer's service data.

Note. After correctly setting the purity, turning the receiver upside down will show the effect of the Earth's magnetic field.

Exercise 5.4

The value of CRT tester/reactivators can be demonstrated if CRTs with low emission and/or inter-electrode short circuits are available. Most testers are capable of clearing many short circuits by discharging a capacitor across the leakage path. In addition, a useful extension of CRT lifetime can often be gained by over-running the cathodes under controlled conditions. This technique can also be used to balance the efficiency of the three colour guns in an older tube.

Test questions

1 Explain with the aid of a sketch why certain types of in-line gun CRT require convergence adjustment.

2 Describe four important safety precautions to be observed when changing a CRT.

3 Explain (with the aid of a sketch where necessary) the functions performed by the CRT rim band.

4 With reference to Figure 5.9, state the typical base-pin voltages to be expected when the CRT is displaying a standard colour-bar test signal, at pins 1, 5, and 7; pins 6, 8 and 11.

5 State the visible effect on the colours displayed by the CRT in Figure 5.9, if R625 is open circuit.

6 Refer to Figure 5.9.
  (a) What would be the visible effect if an internal short circuit existed between pins 5 and 8?
  (b) A picture is displayed with the white areas yellow, and the blue areas black. What are the likely causes?

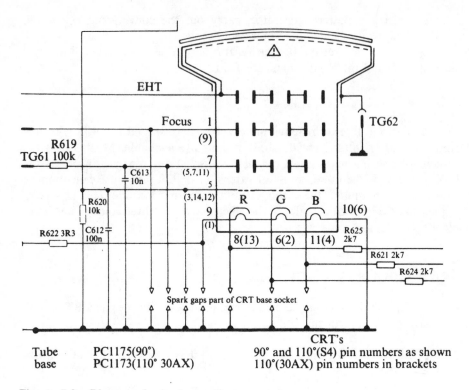

Tube	PC1175(90°)	CRT's
base	PC1173(110° 30AX)	90° and 110°(S4) pin numbers as shown
		110°(30AX) pin numbers in brackets

Figure 5.9   Diagram for test questions

6   Audio stages and power supplies

Summary

Power amplifiers. Stereo systems. Heat sinks. Thermal resistance. Biasing. Class of operation. Setting bias levels. Impedance matching. Feedback. Power-supply types. SMPS. Mains pollution. ICC 5 power supply. Stand-by mode. Soft start. Fault finding.

Power amplifiers

Most current TV receivers use IF demodulator/sound processing ICs that either produce a relatively low power output (up to about 4 W), or provide a small-signal audio output suitable for driving power amplifiers constructed from discrete components in the manner shown in Figure 6.1. Typically such stages will provide output powers up to about 5 W maximum into an 8Ω load. Where higher power is required, the output transistors may be replaced by Darlington Pair devices.

For TV stereo applications where the receiver is equipped to handle the European Standard NICAM 728 (Near Instantaneous Companded Audio Multiplex—728 kb/s data rate) audio system, the analogue stages will be designed to handle the wide frequency range while generating very low levels of noise and distortion. Such receivers will therefore be equipped with two identical audio channels, which include not only volume and tone controls, but also one for stereo balance. In this system, the left and right audio channels are

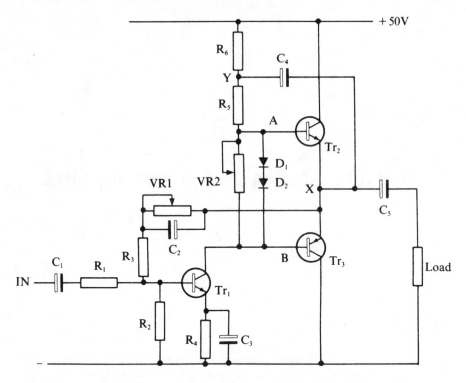

Figure 6.1   Complementary push–pull audio output stage

interleaved into a single bitstream that is digitally modulated on to a subcarrier 6.552 MHz above the vision carrier (IF = 32.948 MHz). The higher-power amplifiers for stereo may provide up to 20 W per channel. Although such a level is unlikely to be needed in a domestic situation, the amplifiers can be operated at a very much lower power level, when they will produce a noise and distortion performance that is commensurate with such a high-fidelity transmission.

Basically, a power amplifier has to handle signals of relatively high voltage and current levels simultaneously. Such devices should therefore possess very low output impedances because the high load current will generate I^2R heat losses that can damage the transistors. Low output impedance is usually achieved by manufacturing transistors with relatively large junction areas. This, together with the use of *heat sinks*, minimizes the risk of permanent damage.

Applications of heat sinks

Heat sinks take the form of finned metal clips or blocks with a large surface area to conduct the heat generated within a transistor to the surrounding air. Good contact between the case of the transistor and the heat sink is essential, and a silicone grease is useful in promoting this contact. Since many types of power transistor have their cases connected to the collector electrode, it is necessary to include an electrically-insulating, heat-conducting mica washer between transistor and heat sink. For good heat conductivity, the washer should be lightly coated with heat-sink grease on both sides. Heat flow through such a system behaves very much like electrical current through series resistances. In fact, a good heat conductor has a parameter described as a low *thermal resistance*, so that an analogue can be drawn as shown in Figure 6.2. The heat-dissipating

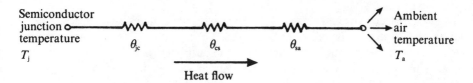

Figure 6.2 Dissipation at heat sink

properties can be explained as follows.

$$\text{Power dissipated, } W = \frac{\text{Temperature drop across system}}{\text{Sum of thermal resistances}},$$

or

$$W = \frac{T_j - T_a}{\theta_{jc} + \theta_{cs} + \theta_{sa}}$$

where θ_{jc} = thermal resistance, junction to case, °C/W; θ_{cs}, thermal resistance, case to sink, °C/W; θ_{sa} = thermal resistance, sink to air, °C/W; T_j = junction temperature; and T_a = ambient air temperature.

This formula can be rearranged to give the maximum value of the thermal resistance for *sink to air* as follows:

$$\theta_{sa} = \frac{T_j - T_a}{W} - (\theta_{jc} + \theta_{cs})$$

Example

A semiconductor device is required to operate with a junction temperature of 100 °C maximum. Its thermal resistance (junction to case) is given as 2 °C/W, and it has to deliver a maximum of 20 W with an ambient air temperature of 20 °C. Assume that the thermal resistance (case to sink), including the mica washer, has a value of 0.5 °C/W.

121

$$\begin{aligned}
\theta_{sa} &= \frac{100 - 20}{20} - (2 + 0.5) \\
&= 4 - 2.5 \\
&= 1.5\,°C/W
\end{aligned}$$

Therefore the heat sink must have a thermal resistance not greater than 1.5 °C/W.

Power-amplifier biasing

To a large extent, the efficiency of the power-amplifier stages is controlled by the class of bias in use. It will be recalled that operating in Class A produces the least amount of distortion but that, since the output current flows for the whole of the input signal cycle (even with no input signal), it is the least efficient (theoretically 50%, with practical values up to 30%). For Class B, where the output current flows only for half the input cycle (it is zero with no input), the efficiency is higher (theoretically 78.4%, with practical values up to 60%). Therefore Class B-biased output stages have to dissipate less wasted heat. Offset against this, Class B push–pull stages suffer from *cross-over* distortion and additionally impose a significant varying load upon the power supply. The typical useful compromise is thus operation in Class AB.

Output impedance matching

Ideally a transformer would be used to match the power-amplifier output impedance into the load impedance of the loudspeaker. However, this device is not only bulky and less than 100% efficient, it can also impose a frequency-response penalty due to its tendency to produce odd resonances. Fortunately, many power-transistor output impedances are low enough to provide a good match directly into an 8 Ω loudspeaker.

Practical output stage

Figure 6.1 shows a typical power output circuit using the totem-pole configuration with Tr_2 and Tr_3 acting in the complementary push–pull mode. Acting as emitter-followers, these two transistors provide an output impedance that can directly match the load impedance. Capacitive coupling is provided by C_5 to avoid the loudspeaker's short-circuiting the d.c. conditions at point X. The circuit operates in Class AB and is capable of providing an output in excess of 10 W with very little distortion.

    The circuit uses both a.c. and d.c. feedback loops. A.c. negative feedback loop is provided by C_2 to counter cross-over distortion and improve the frequency response. A d.c. negative feedback loop is provided via VR1 and R_3 from point X to stabilize the bias current of Tr_1. Postive feedback is taken to point Y via C_4, a technique sometimes referred to as *bootstrapping*. This cannot produce oscillations because, with an emitter-follower, the output signal voltage is always marginally less than that at the input. Since C_4 has a relatively large value, the d.c. voltage across it cannot change very quickly. If the voltage at point X drifts high through a temperature change, the voltage at point Y will also be carried high by the charge on C_4. Since both emitter and base voltages have risen by about the same value, the base bias on Tr_2 is unchanged. The diodes D_1 and D_2 are used to provide the base of Tr_3 with a bias voltage that is about 1.2 V below that of Tr_2. Since the two diodes are made from the same semiconductor material (silicon) as the output transistors, they have a matching temperature characteristic and so stabilize the base voltage at point B.

Two bias-setting adjustments are provided. VR1 sets the bias level of Tr_1 and hence its collector current, which in turn controls the emitter currents of Tr_2 and Tr_3. This control is therefore set so that the d.c. voltage at point X is exactly half that of the supply rail value. If a d.c. multi-range meter is used for this operation, it is important to recognize that such a meter will shunt one of the transistors and give a false reading. Unless a high-impedance meter is available, this problem can be avoided by using two meters connected one across each transistor. VR1 is then set so that both meters read the same value. VR2 is adjusted so that both transistors are not cut off together, and to set the quiescent emitter/collector currents to some specified value. This is achieved by breaking the d.c. collector feed to one of the transistors and inserting a current meter. Typical current levels will range from about 2 to 20 mA depending upon the level of output power available from the two transistors.

Power supplies

Power supplies were fairly extensively covered in Volume 2, Part 1, Chapter 5 of this series and the reader is therefore referred to this for any necessary revision. The circuits included in this section are used to exemplify some of the techniques that might be found in a practical receiver.

TV receiver power-supply techniques vary considerably, ranging from the traditional simple multi-winding transformer driven systems, to complex *switched mode power supplies* (SMPS) linked to the line timebase frequency. Many also use bridge-rectifier assemblies directly connected to the mains supplies and this introduces further problems for the service engineer. The *earthy* side of the receiver is now at half mains-supply potential instead of the expected true earth level. Using any mains-driven test equipment with a direct earth connection leads to a massive short-circuit that can produce disastrous results. Such a problem can only be safely solved by using a mains *isolation transformer*.

Many of the sub-systems in a TV receiver have a large current-varying demand and this leads to the need for some form of stabilization. For the conventional power supply, the series or shunt regulator transistor operates in the linear mode, and these circuits seldom have an efficiency greater than about 30%. Since the wasted power has to be dissipated as heat, such circuits need to mounted on relatively large heat-sinks. By comparison, SMPS transistors operate in a switching mode and are therefore either in the saturated or cut-off state. This leads to power efficiencies as high as 85%, so that heat-dissipation problems are minimal. Higher-frequency operation means that the ripple frequency is also higher, so that the ripple filter can be both smaller and more efficient. Further, the size of any mains-input transformer tends to be inversely proportional to the frequency of operation. Thus if an SMPS operates at 15.625 kHz, the unit can be very much smaller. In addition, SMPS systems can easily be adapted for battery operation, and this allows a mains-driven receiver to be used in a portable manner.

Interference and mains pollution

The use of high-frequency switched-mode power supplies greatly increases the problems associated with the radiation and transmission of radio-frequency interference (RFI), both within the receiver and along the power-supply lines to create pollution for other users. The small transformer shown in Figure 6.3

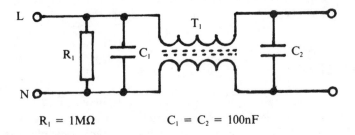

$R_1 = 1M\Omega$ $\qquad\qquad C_1 = C_2 = 100nF$

Figure 6.3 Mains interference filter

consists of two windings on a common core, phased so that there is no coupling between the winding at 50 Hz. The two windings together with the shunt capacitors form a low-pass filter at the mains frequency. At higher frequencies, the interference due to the SMPS is prevented from escaping into the mains-distribution system. The small circuit also provides a useful barrier against common-mode interference signals that may *see* the power lead as an aerial.

A practical TV power supply

The circuit diagram shown in Figure 6.4 forms part of the Ferguson ICC 5 chassis which has its origins in Europe, having been used under the various badges of the French Thomson organization. The major section of the power supply has its mains isolation provided by the *chopper* and pulse transformers LP32 and LP04. The shunt-type chopper transistor TP24 is driven by a complementary symmetry pair TP16 and TP19, via TP11 and transformer LP32 with line-frequency pulses from the power processor chip.

At start-up, the driver stages draw power from the centre tap in the mains transformer, with diodes DP08 and DP09 acting as rectifiers. Under normal operation, power for the chopper circuit is provided from the transformer winding 8,9 and DP26. A *soft-start* feature is provided so that components are not stressed during the start-up period. At power-up, the power processor chip IL14 (Figure 6.5) provides a pulse stream with a short mark-to-space ratio. This progressively increases so that the output voltages rise slowly. Four auxiliary d.c. supplies, half-wave rectified, provide power for the signal and timebase circuits. Because of the high operating frequency, these circuits need only relatively simple ripple filters. Transistors TP14 and TP15 in Figure 6.4 provide an excess-current trip circuit. If the chopper-transistor current rises above some predetermined level, the voltage across RP21 and RP25 causes TP15 to switch on. This creates a base current in TP14 which starts to conduct with its collector current providing further drive for TP15 base. The circuit now latches on and DP16 becomes conductive so that TP15 short-circuits the pulse stream provided by TP11.

As with many modern receivers, this chassis provides for *stand-by* operation, and Figure 6.5 indicates the way in which this is achieved. Two conditions can be identified. When the main power switch is off, the receiver is completely dead; however, if turned off via a remote control signal, the set goes into the *stand-by* mode. This causes a flip-flop in the remote-control chip IR01 to switch to a low output. If the stand-by 5 V supply is present, TR26 and TP48 turn on to provide a 12 V supply for the power processor chip IL14, from the separate rectifier and regulator supply (TP45). If the 5 V stand-by voltage is absent, the set will be dead.

The power-processor chip IL14 (TEA 2029C) not only provides line and field timebase drives, and the pulse-width modulator drive for the SMPS, but also the soft-start and safety-shutdown features to provide a comprehensive protection facility. These features are shown in Figure 6.6 which represents part of the processor chip. The input to IL14 (pin 9) monitors the HT and 36 V lines and compares these samples with a reference voltage. These two voltages are monitored by the networks RL10, PL15 and RP50, RP51 and RL25 respectively. This op-amp provides an output that changes in opposite polarity to any changes at its input to set the switching point of the pulse-width modulator.

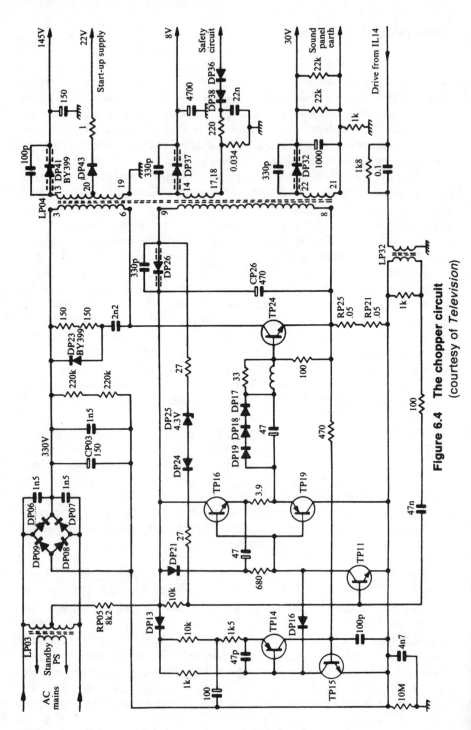

Figure 6.4 The chopper circuit (courtesy of *Television*)

126

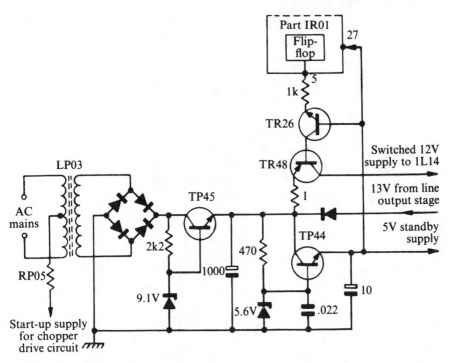

Figure 6.5 **Stand-by/start-up circuit**
(courtesy of *Television*)

This part of the circuit is also driven by a line-frequency sawtooth voltage. The chopper circuit drive output is provided via pin 7. The soft-start feature is controlled by the charge on CL21 connected to pin 15. As this discharges, the op-amp output falls steadily to its normal level, gradually increasing the output voltage. Two pins (9 and 19), not shown on the diagram, represent very high input-impedance inputs and these should not be shunted by using meter probes. Probing pin 9 will cause an instant power failure, and any extra loading on pin 19 will destroy the chopper transistor. If these points have to be monitored during servicing, then an oscilloscope can be connected, but only if the receiver is first switched off.

Fault finding

Servicing in audio stages and power supplies requires two distinctly different approaches. While the actual method adopted is a matter of personal prefer-ence, the former lends itself to either the signal-tracing or signal-injection technique, using either the end-to-end or half-split methods. SMPS, with their

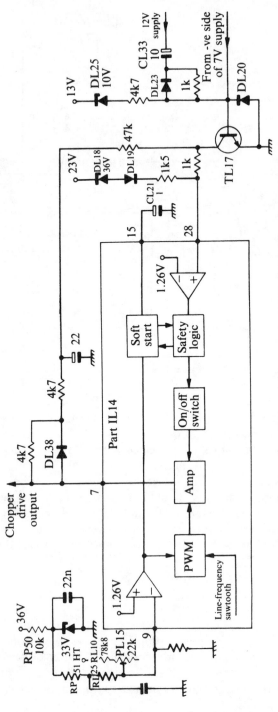

Figure 6.6 Monitoring and protection part of TEA 2029C power processor (courtesy of *Television*)

multitude of feedback and control loops, make it necessary to take an overall view of all symptoms and the various voltage levels. Used with care, an oscilloscope can be a valuable asset when fault-finding in these circuits.

Exercises and test questions

For Chapter 6, these two features have been combined. Either of the circuits involved here can easily be constructed on a breadboard or more permanently on a PCB. The breadboard method is recommended because it readily allows open- or short-circuit component failures to be simulated. Both circuits contain feedback and/or control loops, and their behaviour under fault conditions should be carefully studied.

Exercise and test questions 6.1 (Figure 6.7)

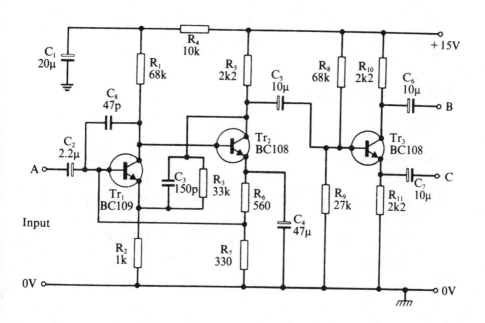

Figure 6.7   Audio amplifier circuit for Exercise 6.1

129

This three-stage audio amplifier provides push–pull outputs at about 2.5 V peak when driven by an input signal of 100 mV peak at the standard sine-wave audio test frequency of 1 kHz.

1   Under normal working conditions the output signal at B has an amplitude of −2.5 V peak relative to the input signal. State, with reasons, the amplitudes and phases of the signals expected at output C and the collector of Tr_2.

2   Explain the effect on the output signals if R_8 becomes open-circuit.

3   Explain the effect on the output signals if R_9 becomes open-circuit.

4   Table 6.1 lists the d.c. voltages obtained with a meter of 20kΩ/V sensitivity under normal and fault conditions. State with reasons which single component could give rise to the no-output condition.

Table 6.1

	Tr_1			Tr_2			Tr_3		
	b	e	c	b	e	c	b	e	c
Normal	1.0	0.4	3.2	3.2	2.6	8.0	4.1	3.6	11.2
Fault 1	1.0	0.4	3.2	3.2	2.6	8.0	4.1	3.6	11.2
Fault 2	1.0	0.4	3.2	3.2	2.6	8.0	0.45	0.45	14.0
Fault 3	0.7	0.2	0.7	0.7	0.4	13.5	4.1	3.6	11.2

Exercise and test questions 6.2 (Figure 6.8)

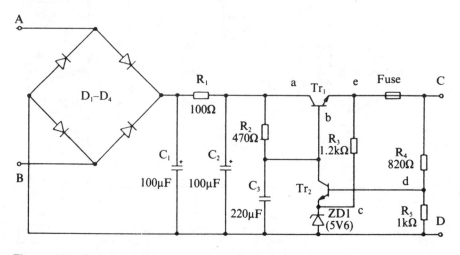

Figure 6.8   Power-supply circuit for Exercise 6.2

The circuit shown is that of a low-voltage stabilized power supply driven from a bridge rectifier and using a series regulator.

1 Explain the effects on the output voltage at CD if one of the rectifier diodes in the bridge becomes open-circuit.

2 Explain the action of the circuit if the output voltage at CD tends to rise.

3 Explain the effects that would occur if R_4 became open-circuit.

4 Table 6.2 gives the voltages taken on a meter with a sensitivity of $20k\Omega/V$ under normal and fault condition. State with reasons which single component could give rise to this condition.

Table 6.2

	Test points				
	a	b	c	d	e
Normal	18.0	12.0	5.6	6.2	11.25
Fault 1	19.5	0	0	0	0
Fault 2	19.5	19.5	0	0	0
Fault 3	19.5	2.8	0	0.7	2.2

Appendix 1   Answers to test questions

Chapter 1   Radio and television receivers

1   (a) $\frac{\lambda}{2} = (3 \times 10^8)/2f = (3 \times 10^8)/(2 \times 600 \times 10^6)$

$= 3/12 = 0.25\,\mathrm{m}$

(b) The vertical elements then lie in the same plane as the electric field of radiation and at right angles to the plane of the electromagnetic field. This then induces an e.m.f. at the signal frequency into the aerial feeder.

2   (a) $10.7\,\mathrm{MHz}$;   (b) $92.3 + 10.7 = 103\,\mathrm{MHz}$;
(c) $92.3 \pm 10.7\,\mathrm{MHz} = 103$ or $81.6\,\mathrm{MHz}$;
(d) $81.6\,\mathrm{MHz}$;   (e) See Figure A1.1.

3   See Figure 1.17.

4   (a) $8\,\mathrm{MHz}$;   (b) $2\,\mathrm{MHz}$;   (c) $6\,\mathrm{MHz}$.

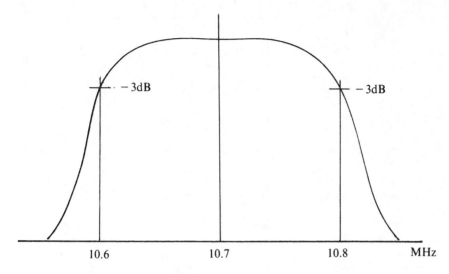

Figure A1.1   Answer to Chapter 1, question 2(e)

Chapter 2   Television fundamentals

1   (a)   4.433 618 75 MHz;
    (b)   detecting or identifying the V line signal component and synchronizing the locally generated subcarrier which was suppressed at the transmitter (DSSC);
    (c) 10 ± 1 cycles.

2   Refer to Figure 2.7 and associated text.

3   (a)   To prevent overmodulation.
    (b)   Using DSSC modulation there is no carrier present when the chrominance signal is zero, as with a monochrome transmission, thus saving on transmitter power.

4   See Figure A1.2. The periodic time is in the order of 52 µs; the remaining 12µs is used by the sync-pulse period.

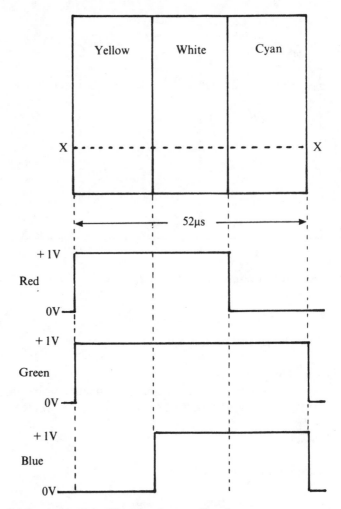

Figure A1.2 Answer to Chapter 2, question 4

Chapter 3 Television circuits

1 (a) Luminance and chrominance delay lines.
  (b) Luminance 700 ns, chrominance 64 μs.
  (c) If the chrominance delay line is o/c the display will only be in mono-chrome. If the luminance delay line is o/c, the white and grey picture areas will be black and the chrominance display will have fuzzy edges. This is due to the restricted bandwidth of the chrominance signal.
  (d) The luminance delay line is used to balance the time delay of the two signals through their respective amplifiers. The chrominance delay line

is used to provide one line delay so that the colour decoder can add and subtract signals from successive lines to separate the U and V components.

2 (a) 4.433 618 75 MHz; (c) 5.5 MHz;
   (b) subcarrier frequency $\pm$ 1 MHz; (d) 5.5 MHz.

3 (a) To remove chrominance information from the luminance channel.
   (b) To prevent the intercarrier 6 MHz sound signal from producing false luminance information.
   (c) False colours.

4 (a) The $G-Y$ colour-difference signal is not transmitted because it is contained within the Y, $R-Y$ and $B-Y$ components. By amplitude scaling, adding and subtracting these components in a colour-difference matrix, the $G-Y$ component can be recovered. The RGB colour signals are obtained by adding the Y signal to the $R-Y$, $G-Y$ and $B-Y$ signals.
   (b) To prevent flyback lines from being displayed on the raster during the retrace periods.

Chapter 4   Time bases and synchronizing circuits

1 Nominally 4.7 µs; 27.3 µs; 2.35 µs; 1.55 µs; 5.8 µs.

2 (a) Critical safety component that should only be replaced by one that meets the appropriate safety standard.
   (b) RV15 and RV5.
   (c) To provide a low impedance load for the field output amplifiers.
   (d) A is the linearity feedback loop to the field-oscillator stage. B is the field pulse drive to IC6, pins 1 and 3.
   (e) Sawtooth voltage.

3 (a) C137.
   (b) L19 and L18 respectively.
   (c) Energy-recovery or efficiency diode. In conjunction with C134 this provides the early part of the line scan.
   (d) Line pulses used for reference in other parts of the circuit, such as phase comparator in the line-oscillator stage and SMPS drive.
   (e) The S-correction will be impaired.

4 (a) Sync separator. (d) S-correction.
   (b) EHT feed to the CRT final anode. (e) Line oscillator.
   (c) U10. (f) R710.

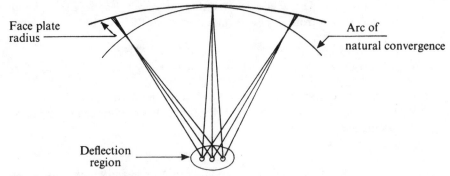

Figure A1.3   Convergence errors

Chapter 5   Cathode ray tubes, associated circuits and liquid crystal displays

1   Due to the difference in radius between the tube face and the natural deflection, as indicated in Figure A1.3, the beams, even if correctly converged near the centre of the screen, will tend to misconverge towards the edges.

2   (a)   Safety goggles and gloves should be worn.
   (b)   Receiver should be switched off, disconnected from the mains and all associated capacitances should be discharged.
   (c)   The tube should not be handled by the neck.
   (d)   The new tube should be retained in its case until needed and the displaced one should be suitably packed as soon as it is removed.

3   Apart from providing support for the CRT in the receiver and a common earth point, the rim band most importantly provides restraint for the forces on the tube bulb that arise due to the air pressure on the tube face. These forces are shown in Figure A1.4.

4   Pin 1, 4–6 kV; pin 5, near-zero; pin 7, 350V.
   Pin 6, 90–100 V; pin 8, 90–100 V; pin 11, 90–100 V.

5   With the loss of red drive signals, the colours will change in the following way:
   white areas turn cyan;
   yellow areas turn green;
   magenta areas turn blue; and
   red areas turn black.
   All other primary and secondary colours will be unaffected.

6   (a)   Grid-to-cathode shorts introduce uncontrollable brightness; therefore images will take on a strong green cast.
   (b)   This fault indicates that the blue signal is absent. This can be caused by a faulty blue-drive amplifier or signal path, faulty blue matrix, or (in

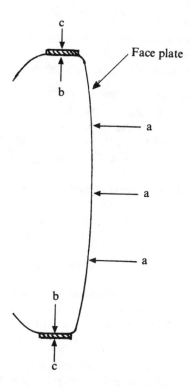

Figure A1.4 **Rim-band: (a) air pressure; (b) outward pressure due to a: (c) rim-band clamping force**

tubes with separate A1 electrodes to each gun) the A1 voltage might be absent.

Chapter 6   Audio stages and power supplies

Exercise 6.1

1   The voltage gain of stage Tr_3 can be calculated from the ratio of collector to emitter resistor. This stage therefore has unity gain at the collector with 180° of phase shift. The output at C must therefore be in phase with the input signal (emitter-follower action), or $+2.5$ V peak. Since Tr_3 has unity gain, this must also be the amplitude and phase of the signal at the collector of Tr_2.

2   If R8 is open-circuit, Tr_3 will be biased to cut-off, so that both outputs will be heavily distorted (half-wave rectification).

137

3 If R_9 is open-circuit, Tr_3 will be biased by the voltage drop across R_8, but will not operate over the linear part of its characteristic. Output B will therefore be distorted. However, as far as output C is concerned, Tr_3 is behaving as an emitter-follower, so that this output is likely to be much less distorted.

4 *Fault 1.* Since the d.c. voltages are unchanged, the fault must be in the signal or a.c. part of the circuit. Thus either C_2 or C_5 must be open-circuit.
Fault 2. Tr_1 and Tr_2 voltages are correct, so the fault must lie with Tr_3. Since the base and emitter voltages are equal, the transistor is cut-off, accounting for the higher collector voltage. Thus an emitter-to-base short-circuit is indicated.
Fault 3. The voltages at Tr_1 collector and base and Tr_2 base are equal, so that the latter is cut-off, and this accounts for its higher collector voltage. The condition can be caused either by an internal short-circuit between collector and base of Tr_1 or a short-circuit in C_8.

Exercise 6.2

1 If one diode is open-circuit, rectification converts to half-wave. The output voltage therefore falls by about 50% and the ripple frequency will fall. The ripple amplitude will thus increase because of the lower efficiency of the filter.

2 If the output voltage rises , that at the base of Tr_2 will also rise and cause this transistor to draw more current through R_2. The base voltage of Tr_1 therefore falls and this reduces the current flowing through it to lower the output voltage. Stabilization action is produced, because R_4 and R_5 provide a sample of the output voltage to be compared with the reference voltage across ZD1. The difference voltage is then used to control the conduction of Tr_1 via Tr_2.

3 If R_4 is open-circuit, Tr_2 will be cut-off and regulation will fail. Tr_1 will still be forward-biased via R_2 and so will pass excessively high current. The output voltage will rise and the fuse should blow.

4 *Fault 1.* Both Tr_1 and Tr_2 are cut off, and zero voltage at test point b indicates either a short-circuit C_3 or an open-circuit R_2.
Fault 2. Because there is no voltage at test points c, d and e, Tr_2 must be cut-off, as must be Tr_1. These suggest that the latter transistor has an internal open circuit between its base and emitter.
Fault 3. The zero voltage at test point c indicates either an open-circuit R_3 or short-circuit zener diode. If R_3 were o/c however, the meter voltage measured at test point c would be non-zero because of the high reverse resistance of the diode. ZD1 must therefore be the faulty component.

Appendix 2   Examination structure

The following examination structure is provided for the award of the Joint Part 2 Certificate in Electronics Servicing of the City and Guilds and the Electronics Examination Boards.

Core studies to be taken by all candidates:
224-2-11   Analogue Electronics Technology (MC)
224-2-12   Analogue Electronics practical assignments (in course)
224-2-13   Digital Electronics Technology (MC)
224-2-14   Digital Electronics practical Assignments (in course)

Options, candidates may enter either one or both of:
224-2-15   Television and Radio Reception Technology (written)
224-2-16   Control System Technology (written)

Practical examination. Candidates may enter one or both, appropriate to the above options:
224-2-17   EEB Practical test in Television and Radio Technology
224-2-18   EEB Practical test in Control System Technology.

The assessments need not all be entered in the same year. The candidate can build up a series of credits towards the ultimate award of the certificate over a period of time.

Appendix 3   Abridged Part 2 Syllabus (index version)

Note: For complete information, the reader is referred to the City and Guilds Part 2 Syllabus.

Part 2

The student is expected to be able to demonstrate competence in setting up, carrying out adjustments and fault finding in either radio and television receivers (colour and monochrome), or basic control systems. He must also demonstrate a general command of basic electrical and electronic principles.

06,   Radio and TV option (Volume 2, Part 2)

6.1   Describe the basic principles of radio and television technology in terms of signal processing through the various sections of the system. Describe the basic concept of colour processing. Describe the structure of the television image (Chapters 1, 2).

6.2   Identify typical sub-sections from a circuit diagram (Chapters 1, 3).

6.3   Describe the construction and operation of the CRT. Recognize the necessary safety precautions required when working on a TV receiver (Chapter 5).

6.4   Describe the luminance and RGB stages of a receiver (Chapter 3).

6.5   Describe the sync-pulse circuitry (Chapter 4).

6.6   Describe the timebase circuits (Chapter 4).

6.7   Describe the waveforms and operating conditions for the time base output stages (Chapter 4).

08,   Science background. Sections 002, 003, 004, 005, 006 (Volume 2, Part 1)

002   Differentiating and integrating circuits.
003   Measuring instruments.
004   Transducers.
005   Electrical principles.
006   Integrated circuits, transformers and displays.

07, Control Systems (Volume 2, Part 3)

7.1   Describe the operation of basic industrial control systems. Explain the terminology associated with the system elements. Explain the purpose of system devices associated with microprocessor-controlled systems.
7.2   State the basic principles of rotating machines, relays, solenoids and actuators.
7.3   Describe the principles and applications of measuring instruments.
7.4   Describe the principles, characteristics and applications of various types of transducers. Describe the operation of circuits incorporating these devices.
7.5   Describe the operation of relays, solenoids and actuators.

08,   Science background

003   Measuring instruments.
004   Transducers.
006   Integrated circuits, transformers and displays.

Appendix 4   EEB practical tests

The test is presented in two sections. The candidate is expected to demonstrate his/her competence to perform a series of analogue and digital measurements using multimeters, double-beam oscilloscopes, logic probes and signal sources. In the second part each candidate is expected to be able to find faults in both analogue and digital circuits to component level.

Measurement tests

The candidate is allowed 45 minutes in which to carry out both static and dynamic measurements on a working circuit, using multimeters and oscilloscope, typically on a simple integrated-circuit colour-bar generator. He/she has also to demonstrate his/her ability to sketch the time-related waveforms found at various test points in the circuit.

Fault finding

The candidate is allowed 2 hours 15 minutes in which to locate three faults in different items of equipment, to component level, and present a comprehensive report on the tests applied, the results found and the conclusions. For the Television and Radio option, the faulty equipment consists of three transistor television receivers from two different manufacturers. For the Control Systems option the equipments are a temperature-control system, a car alarm and an angular position control servo system.

Appendix 5   Assessment checklist (Radio and Television circuits: Section 06)

In a practical situation, each candidate is expected to observe safe working practices and safety precautions.

Use multimeter and oscilloscope to measure and record waveforms and voltages at selected test points in the following stages: luminance demodulator and output; sync separator; field oscillator/output; line oscillator/output.

Use a colour-bar/pattern generator or test card to carry out the following adjustments: A1 and grey scale; focus voltage; picture height and width; EW correction.

Use a signal generator to carry out the RF and IF alignment of an AM radio receiver.

Diagnose faults to component level in separate television receivers.

Appendix 6   Surface-mount component codes

Many of these devices are too small to add the usual identification codes and, for this reason, surface-mount (SM) components should be retained in their packages until required. Devices that are marked usually use a two- or three-character coding as tabulated below:

A	= 1	M	= 3	Y	= 8.2	0 =	$\times 10^0$
B	= 1.1	N	= 3.3	Z	= 9.1	1 =	$\times 10^1$
C	= 1.2	P	= 3.6	a	= 2.5	2 =	$\times 10^2$
D	= 1.3	Q	= 3.9	b	= 3.5	3 =	$\times 10^3$
E	= 1.5	R	= 4.3	d	= 4	4 =	$\times 10^4$
F	= 1.6	S	= 4.7	e	= 4.5	5 =	$\times 10^5$
G	= 1.8	T	= 5.1	f	= 5	6 =	$\times 10^6$
H	= 2	U	= 5.6	m	= 6	7 =	$\times 10^7$
J	= 2.2	V	= 6.2	n	= 7	8 =	$\times 10^8$
K	= 2.4	W	= 6.8	t	= 8	9 =	$\times 10^{-1}$
L	= 2.7	X	= 7.5	y	= 9		

For three-character applications, the first two digits yield the base number and the third the multiplier. For example, 472 = 4.7 kΩ or 4700 pF; 4R7 = 4.7 Ω.

The two-character version consists of a character and a multiplier. For example, A1 = 10 Ω or 10pF; N3 = 3.3 kΩ or 3300 pF.

Appendix 7

POLYGLOT GLOSSARY

English	Danish	Dutch	French
Accumulator	Akkumulator	Accumulator	Accumulateur
Address	Adressere	Adres	Adresse
Amplifier	Forstærker	Versterker	Amplificateur
Antenna	Antenne	Antenne	Antenne
Battery	Batteri	Batterij (or Accu)	Batterie
Cable	Kabel	Kabel	Câble

Capacitors

English	Danish	Dutch	French
Capacitor, fixed, ceramic dielectric	Kondensator, fast, keramisk dielektrisk	Condensator, vaste, mica	Condensateur, fixe, diélectrique en céramique
Capacitor, fixed, electrolytic, aluminium	Kondensator, fast, elektrolytisk, aluminium	Condensator, vaste, aluminium	Condensateur, fixe, électrolytique, aluminium
Capacitor, fixed, electrolytic, tantalum	Kondensator, fast, elektrolytisk, tantal	Condensator, vaste, tantaal	Condensateur, fixe, électrolytique, tantale
Capacitor, fixed, mica dielectric	Kondensator, fast, glimmer dielektrisk	Condensator, vaste, mica	Condensateur, fixe, diélectrique en mica
Capacitor, fixed, paper dielectric	Kondensator, fast, papir dielektrisk	Condensator, vaste, papier	Condensateur, fixe, diélectrique en papier
Capacitor, fixed, plastic dielectric	Kondensator, fast, plastisk dielektrisk	Condensator, vaste, kunststof	Condensateur, fixe, diélectrique en plastique

German	Italian	Swedish
Akkumulator	Accumulatore	Ackumulator
Adresse	Indirizzo	Adress
Verstärker	Amplificatore	Förstärkare
Antenne	Antenna	Antenn
Batterie	Batteria	Batteri
Kabel	Cavo	Kabel
Festkondensator, Keramik-Dielektrikum	Condensatore, fisso, dielettrico in ceramica	Kondensator, fast, keramisk dielektrisk
Festkondensator, elektrolytisch, Aluminium	Condensatore, fisso, elettrolitico alluminio	Kondensator, fast, elektrolytisk, aluminium
Festkondensator, elektrolytisch, Tantal	Condensatore, fisso, elettrolitico, tantalio	Kondensator, fast, elektrolytisk tantal
Festkondensator, Mika-Dielektrikum	Condensatore, fisso, dielettrico a mica	Kondensator, fast, glimmer dielektrisk
Festkondensator, Papier-Dielektrikum	Condensatore, fisso, dielettrico in carta	Kondensator, fast, papper dielektrisk
Festkondensator, Plastik-Dielektrikum	Condensatore, fisso, dielettrico in plastica	Kondensator, fast, plast dielektrisk

English	Danish	Dutch	French
Capacitor, variable air dielectric, differential	Kondensator, variabel, luft dielektrisk differential	Condensator, variabele, lucht diëlectricum, differentiële	Condensateur, variable, diélectrique air, différential
Capacitor, variable air dielectric, trimming	Kondensator, variabel, luft dielektrisk, trimming	Trimmer-condensator, variabele, lucht diëlectricum	Condensateur, variable, diélectrique air, trimmer
Capacitor, variable air dielectric, tuning	Kondensator, variabel, luft dielektrisk, indstilling	Afstem-condensator, variabele, lucht diëlectricum	Condensateur, variable, diélectrique air, d'accord
Capacitor, variable ceramic dielectric, trimming	Kondensator, variabel, keramisk dielektrisk, trimming	Trimmer-condensator, variabele, keramisch diëlectricum	Condensateur, variable, diélectrique en céramique, trimmer
Capacitor, variable mica dielectric, trimming	Kondensator, variabel, glimmer dielektrisk, trimming	Trimmer-condensator, variabele, mica diëlectricum	Condensateur, variable, diélectrique en mica, trimmer
Capacitor, variable, plastic dielectric, trimming	Kondensator, variabel, plastisk dielektrisk, trimming	Trimmer-condensator, variabele, kunststof diëlectricum	Condensateur, variable, diélectrique en plastique, trimmer
Capacitor, variable, solid dielectric, tuning	Kondensator, variabel, massiv dielektrisk, indstilling	Afstem-condensator, variabele, vaste-stof	Condensateur, variable, diélectrique solide, d'accord
Channel	Kanal	Kanaal	Canal
Clip	Klemme	Klem	Attache

German	Italian	Swedish
Drehkondensator, Luft-Dielektrikum, Differential	Condensatore, variabile, dielettrico aria, differenziale	Kondensator, variabel, luft dielektrisk, differential
Drehkondensator, Luft-Dielektrikum, zum Trimmen	Condensatore, variabile, dielettrico aria, compensazione	Kondensator, variabel, luft dielektrisk, trimning
Drehkondensator, Luft-Dielektrikum, zum Abstimmen	Condensatore, variabile, dielettrico, aria, sintonia	Kondensator, variabel, luft, dielektrisk, inställning
Drehkondensator, Keramik-Dielektrikum, zum Trimmen	Condensatore, variabile, dielettrico in ceramica, compensazione	Kondensator, variabel, keramisk dielektrisk, trimning
Drehkondensator, Mika-Dielektrikum, zum Trimmen	Condensatore, variabile, dielettrico mica, compensazione	Kondensator, variabel, glimmer dielektrisk, trimning
Drehkondensator, Plastik-Dielektrikum, zum Trimmen	Condensatore, variabile, dielettrico in plastica, compensazione	Kondensator, variabel, plast dielektrisk, trimning
Drehkondensator, festes Dielektrikum, zum Abstimmen	Condensatore, variabile, dielettrico solido, sintonia	Kondensator, variabel, solid dielektrisk, inställning
Kanal	Canale	Kanal
Klemme	Fermaglio	Klämma

English	Danish	Dutch	French
Coaxial	Koaksialt	Coaxiale	Coaxial
Coil	Spole	Spoel	Bobine
Connecting strip	Forbindelses-strimmel	Verbindings strip	Lamelle de connexion
Crystal unit	Krystal-aggregat	Kristaleenheid	Ensemble à cristal
Dial	Nummerskive	Afleesschaal wijzerplaat kiesschijf	Cadran
Disc	Skive	Schijf	Disque
Earphones	Hovedtelefon	Hoofdtelefoon	Écouteurs
Field effect transistor	Felt effekt transistor	Veldeffekt transistor	Transistor à effet de champ
Frequency	Frekvens	Frequent	Fréquence
Fuse	Sikring	Zekering	Fusible
Fuse holder	Sikersholder	Zekering houder	Porte-fusible
Group board	Grupperings-tavle	Verdeelbord	Tableau de distribution
Hermetic seal	Hermetisk forsegling	Afdichting	Joint hermétique
Input	Indgangseffekt	Invoer	Entrée
Insulator, stand-off	Isolator, afstand	Isolator, afstand	Colonne isolante
Jack plug and socket	Jack-stikkontakt	Telefoonklink met plug	Fiche et douille de jack

150

German	Italian	Swedish
Koaxial	Coassiale	Koaxial
Spule	Bobina	Spole
Klemmleiste	Fascetta di collegamento	Förbindningsbleck
Kristalleinheit	Gruppo cristallo	Kristallenhet
Zifferblatt	Quadrante	Nummerskiva
Scheibe	Disco	Skiva
Kopfhörer	Cuffia radio-telefonica	Hörlurar
Feldeffekt-Transistor (Fieldistor)	Transistore a effetto di campo	Fälteffekt Transistor
Frequenz	Frequenza	Frekvens
Sicherung	Fusibile	Propp
Sicherungshalter	Portafusibile	Propphållare
Anordnungstafel	Quadro del gruppo	Grupperingsbräda
Hermetische Abdichtung	Guarnizione ermetica	Hermetisk försegling
Eingabe	Entrata	Ineffekt
Isolator, Distanz	Isolatore, portante	Isolator, standoff
Klinkenstöpsel und Steckdose	Spina singola e presa	Jackpropp och uttag

English	Danish	Dutch	French
Knob	Knap	Knop	Bouton
Lamp	Lampe	Lamp	Lampe
Lamp holder	Lampeholder	Lamphouder	Douille de lampe
Level	Niveau	Nivo	Niveau
Loudspeaker	Højttaler	Luidspreker	Haut-parleur
Meter	Måler	Meter	Enregistreur
Microphone	Mikrofon	Microfoon	Microphone
Motor	Motor	Motor	Moteur
Output	Udganseffekt	Uitgangs-vermogen	Sortie
Plug and socket	Stikkontakt	Steker en stekerbus	Prises mâle et femelle
Printed circuit	Trykt Kredsløb	Gedrukte stroomkring	Circuit imprimé
Printed wiring connector	Trykt Ledningsfor-bindelse	Printplaat connector	Connecteur de câblage imprimé
Push-button switch	Trykkontakt	Drukknop schakelaar	Commutateur à bouton-poussoir
Radio frequency	Radiofrekvens	Hogefrequentie	Haute fréquence
Rectifier, metal	Ensretter, metal	Metaal gelijkrichter	Redresseur sec

German	Italian	Swedish
Knopf	Pomella	Knapp
Leuchtkörper	Lampada	Lampa
Leuchtkörperfassung	Portalampada	Lamphållare
Pegel	Livella	Nivå
Lautsprecher	Altoparlante	Högtalare
Zähler	Strumento di misurazione	Mätare
Mikrofon	Microfono	Mikrofon
Motor	Motore elettrico	Motor
Ausgabe	Uscita	Uteffekt
Stecker und Steckdose	Spina e presa	Stickontakt och utlag
Gedruckte Schaltungen	Circuito stampato	Tryckt kretslopp
Gedruckte Leitungs-verbindungen	Serrafili circuito stampato	Tryckt kabelförbindning
Druckschalter	Interruttore a puesante	Tryckknappsströms-tällare
Radiofrequenz	Radiofrequenza	Radiofrekvens
Gleichrichter, Metall	Raddrizzatore, metallo	Likriktare, metall

English	Danish	Dutch	French
Rectifier, semiconductor	Ensretter halvleder	Halfgeleider gelijkrichter	Redresseur semiconducteur
Rectifier, silicon	Ensretter silicum	Silicium gelijkrichter	Redresseur au silicium
Relay	Relæ	Relais	Relais

Resistors

English	Danish	Dutch	French
Resistor, fixed, carbon composition	Modstand, fast, Kulstofkomposition	Weerstand, vaste, koolmassa	Résistance, fixe, agglomérée de carbone
Resistor, fixed, cracked carbon	Modstand, fast, krak-kulstof	Weerstand, vaste, opgedampte kool	Résistance, fixe, carbone de craquage
Resistor, fixed, metal film	Modstand, fast, metalhinde	Weerstand, vaste, metaalfilm	Résistance, fixe, couche métallique
Resistor, fixed, oxide film	Modstand, fast, oksydhinde	Weerstand, vaste, oxidefilm	Résistance, fixe, couche d'oxyde
Resistor, fixed, wirewound, general purpose	Modstand, fast, trådomviklet, almindeligt formål	Weerstand, vaste, universele, draad	Résistance, fixe, bobinée, universelle
Resistor, fixed, wirewound, precision	Modstand, fast, trådomviklet, præcisions	Weerstand, vaste, precisiedraad	Résistance, fixe, bobinée, de précision
Resistor, variable, carbon	Modstand, variabel, kulstof	Weerstand, variabele, kool	Résistance, variable, carbone

German	**Italian**	**Swedish**
Gleichrichter, Halbleiter	Raddrizzatore, semiconduttore	Likritare, halvledare
Gleichrichter, Silikon	Raddrizzatore, silicio	Likritare, kisel
Relais	Relè	Relä
Festwiderstand, Kohlenstoff-zusammensetzung	Resistenza, fissa, composizione di carbone	Motstånd, fast, kolsammansättning
Festwiderstand, gesprungene Kohle	Resistenza, fissa, frammenti di carbone	Motstånd, fast, fragmenterat kol
Festwiderstand Metallfilm	Resistenza, fissa, pellicola	Motstånd, fast, metallhinna
Festwiderstand, Oxidfilm	Resistenza, fissa, pellicola, ossidata	Motstånd, fast, oxidhinna
Festwiderstand, drahtgewickelt, für Allegemeinzwecke	Resistenza, fissa, avvolgimento in filo, impiego generale	Motstånd, fast, trådlindad, allmänna ändamål
Festwiderstand, drahtgewickelt, Präzision	Resistenza, fissa, avvolgimento in filo, di precisione	Motstånd, fast, trådlindad, precision
Drehwiderstand, Kohle	Resistenza, variabile, carbone	Motstånd, variabel, kol

English	Danish	Dutch	French
Resistor, variable, wirewound, general purpose	Modstand, variabel, trådomviklet, almindeligt formål	Weerstand, variabele, universele draad	Résistance, variable, bobinée, universelle
Resistor, variable, wirewound, precision	Modstand, variabel, trådomviklet, præcisions	Weerstand, variabele, precisiedraad	Résistance, variable, bobinée, de précision

Switches

English	Danish	Dutch	French
Key switch	Nøgle-afbryder	Druktoets-schakelaar	Manipulateur
Microswitch	Mikroafbryder	Microschakelaar	Micro-contact
Rotary switch	Roterende afbryder	Draai schakelaar	Commutateur rotatif
Slide-action switch	Glide-afbryder	Schuif-schakelaar	Commutateur à curseur
Toggle switch	Leddet afbryder	Tuimel-schakelaar	Tumbler
Sleeving	Overtræk	Mantel	Manchonnage
Terminal	Tilslutnings-klemme	Aansluitpunt	Borne
Transistor	Transistor	Transistor	Transistor
Transformer audio	Transformator audio	Transformator audio	Transformateur, audio fréquence
Transformer IF	Transformator mellemfrekvens	Transformator midden frequent	Transformateur, fréquence intermédiaire

German	Italian	Swedish
Drehwiderstand, drahtgewickelt, für Allegemeinzwecke	Resistenza, variabile, avvolgimento in filo, impiego generale	Motstånd, variabel, trådlindad, allmänna ändamål
Drehwiderstand, drahtgewickelt, Prazision	Resistenza, variabile, avvolgimento in filo, di precisione	Motstånd, variabel, trådlindad, precision
Schlüsselschalter	Interruttore a chiave	Nyckelströmställare
Mikroschalter	Microinterruttore	Mikroströmställare
Drehschalter	Interuttore a rotazione	Roterande strömställare
Schiebeschalter	Interrutore scorrevole	Glid-strömställare
Kippschalter	Interruttore a levetta	Ledad strömställare
Umhüllungen	Manicotto	Hylsor
Anschluss	Terminale	Kabelfäste
Transistor	Transistore	Transistor
Transformator niederfrequenz	Trasformatore, audio	Transformator, lågfrekvens
Transformator, Zwischenfrequenz	Trasformatore, frequenza intermedia	Transormator, mellanfrekvens

English	Danish	Dutch	French
Transformer mains	Transformator lysnet	Transformator voedings	Transformateur, d'alimentation
Toroidal coil	Toroid-spole	Toroidespoel	Bobine toroïdale
Valve	Rør	Buis	Tube
Valve holder	Rørholder	Buishouder	Support de tube
Valve screen	Rørskærm	Buisscherm	Ecran de tube
Valve, subminiature	Lille miniaturrør	Subminiatuur buis	Tube subminiature
Wire	Ledning	Draad	Fil
X Plates	X-Afdriftpladen	X-Abuigplaten	Plaques de déviation horizontale
Y Plates	Y-Afdriftpladen	Y-Afbuigplaten	Plaques de déviation verticale

German	Italian	Swedish
Transformatornetz (Netztrafo)	Trasformatore linea principale	Transformator, huvudledning
Toroidspule	Bobina toroidale	Toroidspole
Röhre	Valvola	Rör
Röhrenfassung	Portavalvola	Rörhallare
Röhrengitter	Schermo valvola	Rörfilter
Kleinströhre	Valvola extra piccola	Subminiatyrrör
Draht	Filo	Ledning
X-Ablenkplatten	Piatti di deviazione orizzontale	X-plattor
Y-Ablenkplatten	Piatti di deviazione verticale	Y-plattor

Index